About IFPRI

The International Food Policy Research Institute (IFPRI), a CGIAR Research Center established in 1975, provides research-based policy solutions to sustainably reduce poverty and end hunger and malnutrition. IFPRI's strategic research aims to foster a climate-resilient and sustainable food supply; promote healthy diets and nutrition for all; build inclusive and efficient markets, trade systems, and food industries; transform agricultural and rural economies; and strengthen institutions and governance. Gender is integrated in all the Institute's work. Partnerships, communications, capacity strengthening, and data and knowledge management are essential components to translate IFPRI's research from action to impact. The Institute's regional and country programs play a critical role in responding to demand for food policy research and in delivering holistic support for country-led development. IFPRI collaborates with partners around the world.

About IFPRI's Peer Review Process

IFPRI books are policy-relevant publications based on original and innovative research conducted at IFPRI. All manuscripts submitted for publication as IFPRI books undergo an extensive review procedure that is managed by IFPRI's Publications Review Committee (PRC). Upon submission to the PRC, the manuscript is reviewed by a PRC member. Once the manuscript is considered ready for external review, the PRC submits it to at least two external reviewers who are chosen for their familiarity with the subject matter and the country setting. Upon receipt of these blind external peer reviews, the PRC provides the author with an editorial decision and, when necessary, instructions for revision based on the external reviews. The PRC reassesses the revised manuscript and makes a recommendation regarding publication to the director general of IFPRI. With the director general's approval, the manuscript enters the editorial and production phase to become an IFPRI book.

Disentangling Food Security from Subsistence Agriculture in Malawi

By Todd Benson

A Peer Reviewed Publication

International Food Policy Research Institute
Washington, DC

Library of Congress Control Number: 2021932929

DOI: https://doi.org/10.2499/9780896294059

Recommended citation: Benson, Todd. *Disentangling Food Security from Subsistence Agriculture in Malawi.* Washington, DC: International Food Policy Research Institute, 2021.

International Food Policy Research Institute, 1201 Eye Street NW, Washington, DC 20005, USA Telephone: 202-862-5600

Cover Design: Jason Chow/IFPRI
Book Layout: Joan Stephens/JKS Design
Cover photos: (top) Reuters/Alamy Stock Photo; (bottom-left) Stephen Dorey/Alamy Stock Photo; (bottom-middle) Julian Lott/Alamy Stock Photo; (bottom-right) Melissa Cooperman/IFPRI.

Contents

Tables, Figures, and Boxes

Tables

Figures

Boxes

Foreword

IFPRI researchers have worked in Malawi since 1990, conducting empirical studies on poverty, nutrition, agricultural transformation, and development strategies. The studies have tried to answer important policy questions, using the best empirical methodologies with conceptual frameworks applied to the local context. Todd Benson draws on more than 25 years of experience in Malawi to synthesize several strands of this research about Malawi's food systems and suggests how Malawi could reduce persistent food insecurity by moving away from its reliance on subsistence farming. With stronger agricultural markets, commercially oriented smallholder farmers across Malawi could efficiently produce much of the food needed. By ensuring an accessible supply of food from dependable markets, the increased production would provide many poor subsistence farmers with the opportunity to pursue more remunerative nonfarm employment.

This approach to achieving food security in Malawi will also be of interest to other countries where most households still rely on their own production to meet their food needs—Malawi is not unique in this regard. The book's analytical frameworks and analyses can be replicated in other countries to generate contextually appropriate guidance for ensuring that all citizens are food secure while also transforming agriculture sectors away from low-input, low-output subsistence farming. This volume should motivate further research, both in Malawi and elsewhere, on many of the issues considered here.

The recommendations put forward can be expected to stimulate debate among Malawian stakeholders on the best policies and programs to steadily improve the food security and broader welfare of all Malawians, with the goal of ending persistent food shortages. This book will serve as an excellent source of inspiration and carefully collected information for all those involved in these debates.

Johan Swinnen
Director General, IFPRI

Acknowledgments

This book is the product of a range of research activities on agriculture and food security in Malawi that I have conducted since 1994. In consequence, my sense of obligation to many collaborators for their guidance and hard work, to hosting institutions for motivating much of my research, and to organizations that provided financial support is deep and heartfelt.

For their contributions to joint research that is discussed in some detail in this book, I specifically want to recognize Sophie Nambuzi, Athur Mabiso, Flora Nankhuni, Brent Edelman, and Grace Kumchulesi. Many others in Malawi also helped me to better understand or to recognize the complexity of the issues discussed here, but I cannot comprehensively list them all. I am grateful to all of them. Smart, committed, hardworking, and reliable colleagues are a real gift.

The institutions that I was a part of or with which I worked closely over the years include the Maize Commodity Team at Chitedze Agricultural Research Station, the National Economic Council (now the Ministry of Economic Planning and Development), the Centre for Social Research of the University of Malawi, the National Statistical Office of Malawi, and the Department of Agricultural Planning Services of the Ministry of Agriculture. These Malawian institutions are a credit to the country and its development ambitions. I am proud and grateful for having worked with them and hope I can continue to do so. I also acknowledge the productive and warm relationship that I have had with the Malawi office of the International Food Policy Research Institute (IFPRI) over more than a decade. The IFPRI-Lilongwe team, both researchers and support staff, have contributed significantly to the strong reputation of the Institute globally and, no doubt, will continue to do so.

Without noting the specific projects under which funding was received, the following organizations, through their generosity, played a vital role in facilitating the research discussed in this book: the Rockefeller Foundation; the Ministry of Agriculture of the government of Malawi; the World Bank; the government of Denmark; the Malawi office of the Department for International Development of the United Kingdom (now part of the Foreign, Commonwealth & Development Office); the United States Agency for International Development, through both the Malawi mission and headquarters; and the Malawi offices of the World Food Programme, the Food and Agriculture Organization of the United Nations, and UNICEF.

In preparing the book, a critical input was the two rounds of detailed comments offered by the two anonymous peer reviewers whom IFPRI's Publications Review Committee asked to provide constructive critiques of the book's contents. My appreciation goes to them. Thanks also to the team in IFPRI's Communications and Public Affairs Division for finalizing the book, particularly for the insightful editing of Pamela Stedman-Edwards.

My greatest debt is to Bets, of course, for being a consistent daily channel of grace of many sorts in periods of both joy and challenge over many years, both in Malawi and elsewhere. My thanks to her for everything.

Acronyms

ADMARC	Agricultural Development and Marketing Corporation
APES	Agricultural Production Estimates System
CGE	computable general equilibrium
FAO	Food and Agriculture Organization of the United Nations
FEWS NET	Famine Early Warning Systems Network
FISP	Farm Input Subsidy Program
GDP	gross domestic product
IHS	Integrated Household Survey
IPS	Integrated Production System
mt	metric tons
NAIP	National Agricultural Investment Plan, 2017/2018 to 2022/2023
NAP	National Agriculture Policy
SAFEX	South African Futures Exchange

Executive Summary

Many rural Malawian households cannot reliably obtain enough food to meet their needs, even though most have access to arable land for food production. Historically, rainfed, low-input subsistence production has been the primary means of assuring household food security. But with rising population pressures on a limited land base, Malawian households must increasingly depend on the country's weak markets for an important share of their food. Yet despite the inadequacy of subsistence farming for meeting the food requirements of all Malawians, the country's principal strategic approach to food security at the policy level continues to center on household agricultural production for own consumption. This reflects a vicious circle in which thin markets for maize, Malawi's staple crop, reduce incentives for commercial production of the crop and foster a reliance on own production by households since they cannot rely on these markets for food security—this, in turn, reinforces the pattern of weak markets. Government policies for food security have largely relied on promoting agricultural production, rather than broader policies of economic diversification and growth to promote a rural transformation that would ensure food security by improving access to food for all. A more viable path to food security is needed that goes beyond subsistence production and the agriculture sector alone.

The approach to sustainable food security in Malawi proposed in this book calls for three critical changes:

- Government support for the emergence of a significantly more diversified and commercially oriented agriculture sector. Specialized, high-productivity farming households will reliably supply Malawi's markets with the food and other agricultural products that the country needs.

- Rural economic transformation propelled by the increased agricultural productivity of these farmers. Workers in many households will increasingly find that they can better provide for the needs of their households by pursuing nonfarm economic activities, producing goods and services for their neighbors who continue to farm and for wider markets, rather than by practicing subsistence farming. However, the government must continue to lead efforts to ensure that the dietary needs of all food-insecure households are met.
- Strengthened food markets, which are central to improving food security for all Malawians. Creating more efficient and deeper markets will require predictable government interventions in markets when necessary, policy stances that are supportive of agricultural market traders, and better integration of Malawi's markets into regional markets. With reliable markets, subsistence farming will become less important for food security, and rural households will be able to improve their welfare through commercial farm production or through a shift to lucrative nonfarm work. Greater price predictability in these markets will reduce risks for both producers and consumers of maize in Malawi, reduce incentives for farm households to focus primarily on low-output subsistence production, and enable many more Malawians to confidently pursue nonfarm employment.

This approach to food security is situated within a process of structural transformation of the economy. As Malawian workers increasingly find remunerative work in the industry and service sectors, subsistence farming will no longer be viewed as strategic for assuring the food needs of their households or of the country as a whole.

Chapter 1

INTRODUCTION

Malawi is a food-insecure country. Although many Malawians continue to farm in order to feed their own families, many cannot reliably meet the dietary needs of all their household members. The four quantitative poverty analyses for Malawi done since the late 1990s have shown consistently that between 50 and 55 percent of all Malawians have consumption levels below the poverty line, based on a basic-needs poverty line that is anchored to the recommended calorie requirements of individuals (Malawi, NSO and World Bank 2007b, 2018). These individuals do not have regular access to enough food to meet their dietary needs. Moreover, rural Malawians who are primarily engaged in farming are more likely to have deficient food consumption than the overall population.

Most of the 4 million households in Malawi farm to obtain an important share of the food they consume. Most farm on land that they received by virtue of being members of a local community, and use hand labor, few modern inputs, and traditional rainfed cropping practices. Production to meet their subsistence needs dominates their livelihood and cropping decisions. Less than a quarter of households produce crops principally for commercial sale. Almost no farming households specialize in the commercial production of nonfood crops, relying on the revenue obtained to buy their food. With most crops being rainfed, there is considerable year-to-year variability in production and, consequently, in the food security of households across the country. With increasing regularity, several hundred thousand households annually are identified as being vulnerable to acute food insecurity due to insufficient crop harvests that result from poor seasonal growing conditions (drought, flood, pests) and limited use of inputs, coupled with reliance on shrinking landholdings as the population continues to grow.

This persistent food insecurity at the individual and household levels strongly influences the policies pursued by the government. The legitimacy of any government in Malawi is largely sustained by citizens' perception that it will be responsive to meeting the needs of the food insecure (Harrigan 2003; Chinsinga and Poulton 2014). In consequence, the government intervenes actively to ensure that food is available. This is generally done in a direct manner, including through providing subsidized fertilizer and improved seed for food crops to a significant share of smallholder farming households, restricting the export of food crops, intervening in agricultural markets by imposing a ceiling on rising food prices, providing vulnerable individuals who are able to work with short-term employment building or rehabilitating community assets (food- or cash-for-work programs), and distributing food aid during acute food insecurity crises.

However, these policies, though possibly necessary and certainly politically acceptable in the short run, are unlikely to result in a food-secure Malawi over the longer term. Achieving such a goal will require the agriculture sector to move away from a focus on production for food self-sufficiency at both household and national levels, and toward a market-centered, more concentrated and specialized sector that, nevertheless, will reliably serve the food needs of increasing numbers of Malawians working outside of the sector. This book provides a contextually informed analysis of what changes Malawians might seek in government policies, in farming systems across the country, in agricultural markets, and in the economies of both local communities and Malawi as a whole to ensure that an economic transformation occurs and all Malawians will have reliable access to the food they require.

Achieving Food Security and Rural Economic Transformation in Malawi

For Malawi to be food secure in the longer term will require a more productive, resilient, commercially oriented, and profitable agriculture sector. The view taken here is that a subsistence-oriented agriculture sector is neither going to assure Malawi's food security nor provide the economic growth necessary to propel a structural transformation of the economy, whether at the national level or within communities across the country. Economic transformation will mean that a much larger share of the population sustainably meets their household needs from livelihood activities other than farming.

In consequence, the approach to achieving sustained food security for Malawi over the longer term described in this book involves government-led efforts to develop a significantly more diversified and commercially oriented agriculture sector overall, with specialized, high-productivity farming households supplying markets with the food and other agricultural products that the country needs. Farmers will focus on production of those commodities they can best produce profitably, rather than producing the wider range of farm products that their households consume.

As this transformation accelerates, most rural households will increasingly find that they can better meet their needs by pursing nonfarm economic activities rather than low-productivity, subsistence-oriented farming. Some will find more remunerative livelihoods in urban centers, but many rural households will find it profitable to reduce their involvement in agriculture and to increase specialized production of goods and services for the local market. These goods and services are those that are labor-intensive, require limited capital in their production, and typically are not marketed outside of the local community—construction, building repair, and associated services; transport services; education, health, and other social services; furniture- and handicraft-making; food and beverage processing; and the like. Demand for these goods and services will be driven by the commercially oriented local farmers. As the income of these more productive local farm households grows, their consumption of locally produced goods and services will increase, in turn increasing the incomes of their neighbors who produce them. With sustained returns, many of the households producing such goods and services for the local market will expand their activities to serve wider markets, further increasing their income and, of interest here, better ensuring their access to food.

A critical part of this approach to improving food security for all Malawians will be strengthening food markets across the country—for both producers and consumers. Producers must be able to bring their output to sell in the market confident that they will obtain a remunerative price from a deep pool of traders and processers operating there. Correspondingly, traders should be able to draw upon a number of sources of supply, primarily through increased domestic production but also through wider sourcing from across eastern and southern Africa. And, as the principal measure of the success of such efforts to strengthen markets, consumers must find that they can go to the market at any time, certain that they will be able to purchase all of the food their households needs within an acceptable price range. Without reliable markets, subsistence farming, even at low levels of productivity, will remain the central livelihood strategy for many poor Malawian households, but few will see their welfare improve.

This book primarily focuses on improving access to food and transforming smallholder farming systems, agricultural markets, and the economies of rural communities. However, the broader context for this rural development is one of structural transformation of Malawi's economy. The most effective incentives for the transformations in food, farming, and market systems considered here will emerge from a more vibrant economy in which subsistence farming activities are increasingly seen by most Malawians as a relatively poor option for meeting the welfare needs of their households. A larger dimension of the approach to food security sketched here is that many more Malawians will find remunerative work in the industry and service sectors of the economy and will increasingly feel that engaging in subsistence farming is no longer strategic for assuring the food needs of their households. Although the topic is not treated in depth, a discussion is presented toward the end of the book of the complementary investments needed to propel economic structural transformation in Malawi. This discussion includes the centrality of urban development to sustainable food security.

Scope of the Book

This book seeks to contribute to the achievement of a development vision for Malawi in which all individuals have access to the food they require. Although the text draws on a range of conceptual approaches to examine and address the issues relevant to this vision, it keeps the context-specific problems that Malawi faces at the center. This book chiefly seeks to better frame and seek solutions for Malawi. The set of development problems related to food security that must be addressed arise from specific features of Malawi's food systems, agricultural production patterns, and economic and historical context. However, the analyses presented and the policy recommendations offered are built from a broader conceptual understanding of food security, smallholder agriculture, rural economic development, and structural change in national economies.

The principal audience for this book is expected to be the broad set of engaged Malawian stakeholders seeking to design strategies to motivate social and economic change for achieving enduring food security and for developing agriculture and rural economies across the country so that the basic needs of all are met. Although the book presents a broad range of evidence from detailed objective research on food security, agriculture, and economic transformation in Malawi, it makes an effort to avoid an overly academic style in order to be of interest to a more general but nonetheless socially engaged audience concerned with the issues raised. More practically, assumptions on the informational needs of three specific segments of such an audience guided its development:

- *Stakeholders* engaged in the design and implementation of policy and programs focused on significantly reducing food insecurity in Malawi and on fostering sustainable development in rural communities and in agriculture. Although this segment of the audience for the book includes many more individuals and organizations than the diverse members of the country's Agricultural Sector Working Group and the technical working groups on agriculture and food security that work under it, the book was developed to provide evidence and present policy options for debate by the members of these groups and the organizations they represent.
- *Students* in agricultural policy–focused courses at universities and colleges in Malawi, particularly Chancellor College and the Lilongwe University of Agriculture and Natural Resources.
- *Staff* of donor agencies and international nongovernmental organizations (NGOs) assigned to Malawi with responsibilities for agriculture or food security.

For these groups, this book provides a relatively accessible overview of the problem of food insecurity in Malawi and what agricultural and rural economic development strategies might be advanced to reduce the misery it causes.

However, the scope of the book's examination of the food insecurity challenge in Malawi is necessarily limited. The author has conducted a wide range of policy-relevant research on Malawi, covering topics including soil fertility management by smallholders, quantitative poverty analysis, and spatial patterns of human development and agricultural production, among others. So the book's content largely reflects this research. However, there are two areas of study that are crucial to achieving and sustaining food security, particularly at national level, for which there is limited discussion in this book.

First, the macroeconomic determinants of food security are not examined in any detail—no discussion is provided of the role that the fiscal, monetary, and exchange rate policies of the government of Malawi might play in alleviating food insecurity. Although some attention is paid to trade policies in the context of strengthening markets, this is not done in detail. Readers seeking guidance for how Malawi might more effectively manage these macroeconomic elements to improve food security will need to consult any number of useful, if not Malawi-specific, guides for doing so (for example, Timmer 2015; Díaz-Bonilla 2015).

Second, the political feasibility of the various policy and program recommendations offered is not considered explicitly. The aim has been to provide contextually relevant solutions from a technical standpoint. Close observation of political debates and dynamics, as well as policy processes, in Malawi over many years contributes to the choice of policy and program recommendations presented. Consequently, their political feasibility is implicit in judging them to be worth offering, but it is not a defining feature. However, important barriers related to culture, language, and history prevent outside observers from developing a reliable understanding of how development priorities are set, policy decisions are made, and resources are allocated in Malawi and what determines the intended distribution of the benefits of those decisions among the citizens of the country. Consequently, in the text that follows, although many suggestions are made as to how a vision of a food-secure Malawi might be achieved, the political debates, strategies, and maneuvering through which any of the suggested actions will be realized are not considered closely.

The book's limitations in dealing with the full range of potential development strategies for Malawi's agriculture sector are also important to recognize. Because the book is organized around food security, it has a strong focus on the agriculture sector and its performance. However, as a consequence, food security is the principal performance indicator examined for the sector, not its aggregate economic production (sectoral gross domestic product, or GDP). Accordingly, little attention is paid to agriculture-sector development strategies that, though they may be important in terms of economic value-added or increasing the value of agricultural exports, are unlikely to directly improve the welfare of the many Malawian households that do some farming as part of their livelihoods. For example, specialized high-value agricultural commodity value chains for export or large-scale foreign direct investment in the sector may raise the aggregate value of production from Malawi's agriculture sector, but any food security rationales for promoting such subsectoral strategies are certain to be quite indirect. Moreover, such strategies will not directly contribute to broad transformation of the economies of rural Malawian communities. This is not to say that such agricultural development approaches do not have a place in efforts to transform the agriculture sector in Malawi. However, we should not expect them to contribute directly in a substantial manner to food security and increased economic opportunities for rural households.

Finally, though the book is primarily intended for a Malawian audience, nonetheless, readers with a particular interest in similar development challenges around food insecurity, agricultural development, and economic transformation facing other low-income countries, particularly in Africa, will find insights here that are pertinent to those challenges. Certainly, several of Malawi's neighbors will find the challenges related to food insecurity that Malawi faces and suggestions for how they might best be addressed to be quite relevant to their own development context. More broadly, the many developing countries around the world that are trying to design effective development pathways to improve livelihoods for the large share of their populations that work in low-output, primarily subsistence-focused agriculture will find lessons presented here that can be adapted for their own specific context.

Organization of the Book

Food security at the household level and nationally is the central organizing theme for this book. Accordingly, Chapter 2 provides a conceptual overview of food security and then examines how burdensome food insecurity has been for Malawian households and the country as a whole in recent years. It also reviews the policies that have been advanced to realize sustainable food security in the country. Given the long-reigning policy view in Malawi that considers food security to be almost synonymous with agriculture, Chapter 3 provides a detailed discussion of the technical basis for and the social organization of agricultural production, followed by an overview of recent policy approaches to agricultural development in the country. The predominantly subsistence orientation of most of those who do any farming in Malawi suggests that the country's food markets are insufficiently robust to meet the food needs of all Malawian households at all times. Chapter 4 focuses on agricultural markets, examining the factors underlying their poor performance and recommending actions to improve them, including the role government could play in strengthening these markets so that they assure access to food for more Malawians.

Chapter 5 brings together the discussions of agriculture-sector performance and that of markets to explore an approach to agricultural development and rural economic transformation that focuses on the commercially oriented smallholder farmer. This approach is first examined conceptually before its feasibility is assessed empirically using recent household survey data. It is clear, given the scarcity of commercially oriented smallholder farming households, that such a strategy for agricultural and rural economic development will face serious challenges and that significant impact will take some time to achieve. However, no other approaches show potential to have as broad an impact on transforming livelihoods within rural communities. Taking a broader look, Chapter 6 provides a brief discussion of the development goal of the structural transformation of the Malawian economy. The closing chapter, Chapter 7, summarizes the policy actions identified in the book that are needed to bring about sustainable food security in Malawi at both the national and household levels.

Chapter 2

FOOD SECURITY IN MALAWI

After providing a conceptual overview of food security and how it contributes to improved nutrition, this chapter examines recent food security trends in Malawi at both the aggregate national level and the household and individual levels. This is done by focusing on maize, the principal staple. Because the country depends largely on rainfed domestic production of maize, the amount available annually is quite variable. The incidence of annual food crises affecting more than 10 percent of the population has increased in recent years. Variation in maize prices across the years, though decreasing, remains high, and seasonal price movements are not always predictable. At the household level, more than half of Malawian households have a level of consumption below the poverty line, which is defined by consumption of a basket of basic goods dominated by food. This poverty rate has been relatively unchanged for the past 20 years. At the individual level, although nutritional outcomes in Malawi are improving along several dimensions, food security and dietary factors have not been important in driving these recent changes, but will be critical for achieving any further advances.

The leaders of Malawi have repeatedly demonstrated that they recognize their obligation to meet the food needs of all citizens. Within the public sector, the agriculture ministry has been the line ministry that has had the heaviest responsibility for doing so, particularly through efforts to increase maize production. However, food security is not synonymous with agriculture. Both improving the employment choices of Malawi's workers and strengthening the markets upon which their households rely for their food are critical food security challenges facing the country. Food security policies and the institutions that implement them should be guided by an understanding of how to ensure food security in Malawi that extends the concept beyond agricultural approaches alone.

The Concept of Food Security

The conceptual framework that guides the discussion around food security here is that of the 1996 World Food Summit Plan of Action: "Food security exists when all people, at all times, have physical and economic access to sufficient, safe and nutritious food to meet their dietary needs and food preferences for an active and healthy life" (FAO 1996). This widely accepted definition has two noteworthy points for our purposes.

First, although the World Food Summit Plan of Action notes that food security is a relevant development objective at the national, regional, and global levels, the definition operates at the household and individual levels. General discussions of food-secure and food-insecure countries are aggregations that do not reflect the considerable variations in the food security of households within a country. Moreover, within the households themselves, the degree to which individuals have access to sufficient food may vary by their sex, age, or labor contribution. Consequently, a clearer understanding of the determinants of and risks to food security in a specific context are best achieved through a disaggregated approach.

Second, food production alone does not equal food security. Rather, food security entails reliable access to food and its proper use to meet dietary needs. Four distinct dimensions are central to the attainment of food security:

- **Availability**—Sufficient quantities of appropriate foods from own production, domestic production, or imports are available within reasonable proximity to all individuals.
- **Access**—Individuals have adequate incomes or other resources to buy, barter, or otherwise obtain enough appropriate foods to maintain an adequate diet. This dimension draws from the observations of Sen (1981) on the critical importance of entitlements to food to ensure food security, rather than food availability alone.
- **Stability**—Stability exists in the availability of food supplies and in access to those supplies. This is the temporal and risk-related dimension of food security (Gross et al. 2000).
- **Utilization**—"Food is properly used; proper food processing and storage techniques are employed; adequate knowledge of nutrition and child care techniques exists and is applied; and adequate health and sanitation services exist" (USAID 1992). This dimension reflects dietary intake as an immediate determinant of the nutritional well-being of an individual, with food security as one of several underlying determinants of nutritional status (UNICEF 1990).

It is the second dimension, that of ensuring reliable access to food, that is the food security focus here. The availability dimension has motivated food security policy in Malawi since the colonial period (Vaughan 1985; Chilowa 1998). That conceptualization of how Malawians can be food secure has resulted in a production-based focus in food security strategies. It also has resulted in the view that household own production of food for themselves should be the dominant means by which Malawians are to be adequately fed and that the agriculture sector, under the leadership of the Ministry of Agriculture and Food Security (hereafter Ministry of Agriculture), has the principal responsibility for assuring the country's food security. Government policy statements that touch on food security, including the current master development strategy, the Malawi Growth and Development Strategy III, conceptualize food security primarily as an outcome of sectoral action around food production, rather than an outcome of a coordinated approach across several sectors to significantly improve the access all Malawians have to the food they require.

Although own production will remain a central element in Malawian food systems for decades to come, the view advocated here is that a reliably food-secure Malawi should not be heavily dependent on an agriculture sector organized around rural households engaged in production primarily to meet their own food needs. Rather, a broader policy is required—one that ensures that all Malawians are able to reliably access food, regardless of how it is made available, but with a focus on dependable markets as the principal source of food for all.

Stability in food supplies is closely related to the access dimension. The recurring food insecurity crises that have affected southern and central Malawi over the past several years, though they stem principally from shocks to food production and, hence, food availability, fundamentally reflect inadequate means of access to food beyond own production. Expanding the means by which households in Malawi and the nation as a whole obtain the food they require beyond subsistence production will result in more resilient food systems.

Finally, the quality of food utilization is fundamental to ensuring that the nutritionally vulnerable in Malawi are well nourished and are able to reach their physical and cognitive potential. Although good nutrition is a critical component of human development, this book principally focuses on ensuring that households have reliable access to foods in sufficient quantity and of appropriate quality to meet their nutritional needs, rather than on the quality of diets that they consume. However, improved nutrition is the ultimate goal of every effort to ensure that all Malawians are food secure. Improving access to food is a critical step toward achieving good nutritional status for all, but other factors also matter. It is possible for individuals to have deficient or unbalanced diets in spite of having good access to diverse foods. A person may have reliable access to the components of a healthy diet, but because of poor health or care, ignorance, or personal preferences with negative nutritional implications, he or she may not be able to, or may choose not to, use the food that is available in a nutritionally sound manner.

More important, to significantly and sustainably reduce malnutrition requires attention to a broad range of issues outside of food systems and how those food systems operate. As shown in the UNICEF conceptual framework of the determinants of nutritional status in young children (Figure 2.1), food security is just one of several underlying determinants of improved nutrition. The UNICEF framework shows that nutritional status is immediately determined by the level and quality of dietary intake and the health status of the young child. The quality of these immediate determinants, in turn, is

FIGURE 2.1 UNICEF conceptual framework of the determinants of young child nutritional status

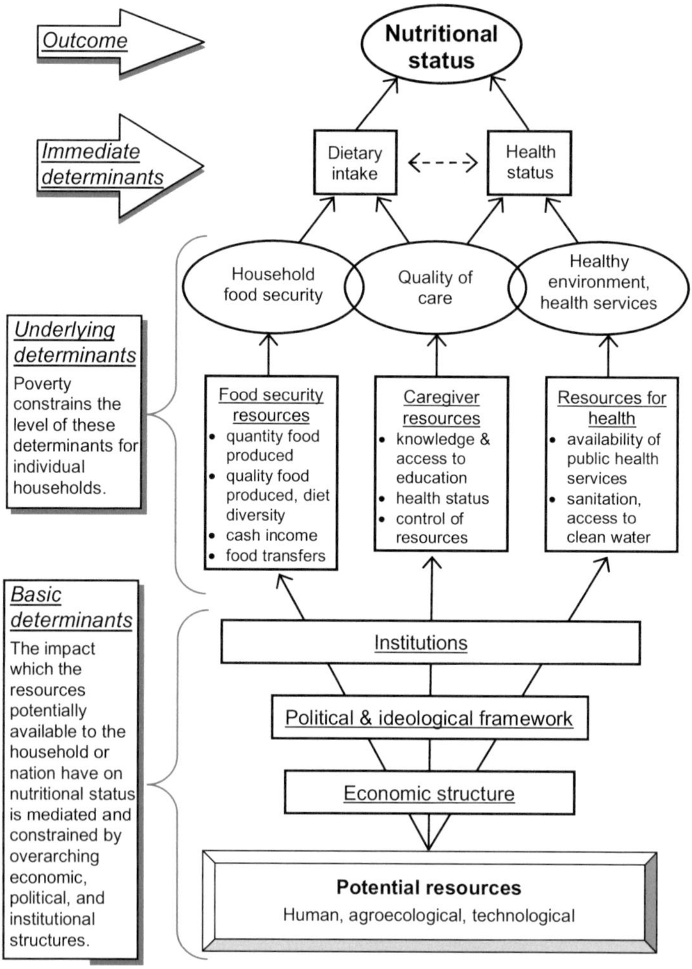

Source: Adapted from UNICEF (1990).

determined by several underlying factors. These include not only food security but also, of equal importance, the availability of health services and a healthy local environment—clean water, good sanitation, adequate housing, and so on—and the quality of care the individual child receives, particularly feeding practices. Food security alone is insufficient to achieve sustained improvements in nutrition. Consequently, food-centered approaches will not be sufficient and food-centered government agencies will not be capable on their own of ending all forms of malnutrition in Malawi. A coordinated approach across multiple sectors of government together with other stakeholders outside of government will be required.

The UNICEF framework also links the availability of nutrition resources to a set of basic determinants. These reflect how society is organized in terms of economic structure, political and ideological expectations, and the institutions through which activities within society are regulated, social values are met, and potential resources are converted into actual resources. Prevailing cultural beliefs and practices around food and the nutritional care of young children are among these basic determinants. Consequently, achieving good nutritional status for all is identified in the framework as a subject for political and social debate and an issue of concern for any national development strategy.

Food Security in Malawi

Food self-sufficiency has been a constant element of agricultural policy in Malawi since the colonial period. The leaders of Malawi, whether colonial governors (Vaughan 1987), the first president of independent Malawi, Kamuzu Banda (Harrigan 2001), or the presidents who succeeded Banda following the advent of multiparty democracy in 1994, all understood that the state they led has an obligation to its citizens to facilitate their access to adequate food. Each successive government has put agricultural activities at the center of its development programs. More recently, nascent social safety net programs have been added to support access to food for vulnerable Malawian households. Moreover, when crop production problems have led to food shortages for vulnerable households, the government has always worked with its development partners to provide those households with food directly.

Within the public sector, the Ministry of Agriculture has been the line ministry with the heaviest responsibility for ensuring the country's food security. Although the narrative has been modified somewhat in recent years, with some attention to the production and consumption of micronutrient-dense crops, for the most part the Ministry of Agriculture has sought to meet the food needs of the population through efforts to sustainably and significantly increase maize productivity. This focus on maize is logical—estimates from the fourth Integrated Household Survey (IHS 4), conducted in 2016/17, are that 74 percent of calories consumed by Malawians come from cereals, primarily maize, with the percentages being somewhat more for rural Malawians, at about 77 percent, and rather less for urban Malawians, at about 65 percent (Gilbert, Benson, and Ecker 2019). No other food group supplies more than 10 percent of calories consumed overall—the second most important food group for calories is pulses, legumes, and nuts, which is estimated to provide 6.5 percent of calories consumed.

Aggregate food security

Consequently, national maize production serves as a suitable measure of food self-sufficiency in Malawi, at least for calorie consumption. Recent national food balance sheet computations assume Malawi's annual maize consumption

FIGURE 2.2 Maize, annual total production for Malawi, average monthly maize prices for Malawi and SAFEX, and maize export ban periods, 2001–2018

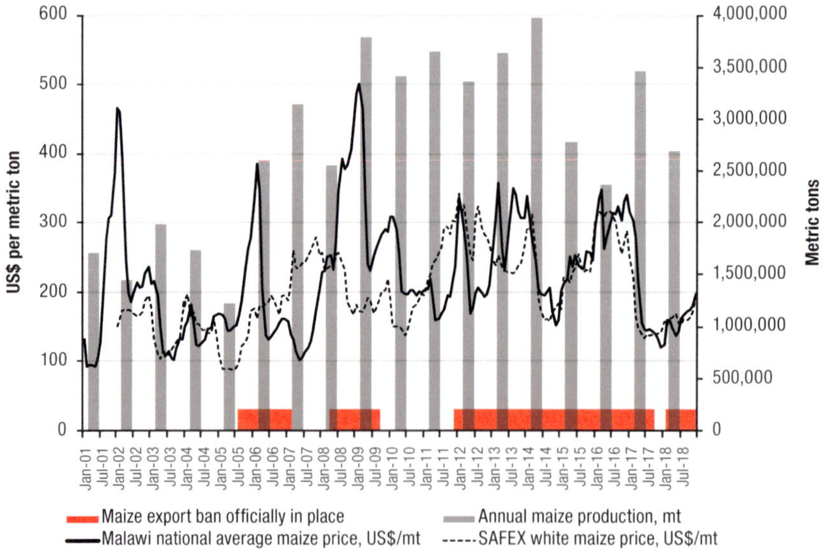

Source: Production: Ministry of Agriculture, Irrigation, and Water Development (MoAIWD) annual production estimates; Prices: MoAIWD to October 2017; International Food Policy Research Institute, Lilongwe, from November 2017; SAFEX data online (https://www.sagis.org.za/safex_historic.html).

Note: SAFEX white maize price series starts January 2002. mt = metric ton; SAFEX = South African Futures Exchange.

requirement is about 2.75 million metric tons,[1] about 150 kg per person per year (Babu et al. 2018). Similarly, maize prices provide a suitable measure of the degree to which food is accessible to all Malawians. Using data series primarily drawn from government sources, annual maize production patterns and monthly average maize prices are shown in Figure 2.2 for the period 2001 to 2018.

The expected relationship between maize production levels and national maize prices—with higher production levels reflected in lower prices and vice versa—is not consistently seen. Although maize prices sometimes spike following years of poor production, then quickly decline once the harvest of the following cropping season starts, for other periods of higher maize prices, this expected pattern is not seen. This inconsistency may reflect deficiencies in production statistics, which falsely suggest sufficient maize production to meet national needs (Box 2.1), political considerations that delay recognition of emerging crises, and often ineffective, poorly managed, or even harmful responses to emerging food crises on the part of government (Ellis and Manda 2012; Chirwa and Chinsinga 2013). Such government responses include placing generally unnecessary export bans on maize (Figure 2.2) and at times directly or indirectly imposing restrictions on private firms' importation or domestic purchase of maize to supply Malawian markets. Many of the government responses to food crises over this period have been made in an unpredictable fashion that raises the level of risk for other participants in maize markets in Malawi and increases the price of maize for all (Pauw and Edelman 2015).

International humanitarian responses to food crises also shape the pattern of maize production and national maize prices. This is most clearly seen in the months following the poor harvest of the 2015/16 cropping season (Figure 2.2). Over this period, the government and its international development partners mobilized the distribution of food to 5.4 million individuals and cash to 1.4 million beneficiaries at a total cost of around US$290 million[2] (Babu et al. 2018). In consequence, what should have been a sharp rise in prices in the months before the 2016/17 harvest was blunted, with prices then falling sharply as the 2016/17 maize was harvested.

We also see in Figure 2.2 that national maize supply in Malawi generally is produced domestically, with limited impact from formal interregional trade. At the aggregate national level, Malawi's maize market remains weakly linked to the southern Africa regional maize market, with small flows of maize under

1 *Tons* refers to metric tons throughout.

2 All dollar amounts are in US dollars.

BOX 2.1 Obtaining accurate agricultural statistics in Malawi: A continuing challenge

Both the Ministry of Agriculture and the National Statistical Office collect information on agriculture in Malawi. The ministry has an annual three-round Agricultural Production Estimates System (APES) for collecting data on crop production and area, and on the number of farmers. The ministry also collects prices for agricultural commodities through the Agriculture Market Information System. The National Statistical Office implements the national Integrated Household Survey (IHS) every three to four years. The IHS questionnaire includes a comprehensive set of modules on agricultural activities and production, as well as on food consumption.

However, the estimates that each agency generates differ significantly or are otherwise inconsistent. For example:

- The 2015 APES estimated that there were 4.12 million farm households nationally. However, the 2018 Malawi Population and Housing Census, conducted by the National Statistical Office three years later, counted 3.33 million rural households and 3.98 million households nationally.
- The 2018/19 APES estimated that total national production of maize was 3.36 million tons; cassava, 5.70 million tons; and sweet potatoes, 6.46 million tons. However, household survey analysis shows that 74 percent of calories consumed by Malawians were from maize (Gilbert, Benson, and Ecker 2019). No other crop was found to supply more than 7 percent. Yet we know that cassava and sweet potatoes are almost exclusively used as food crops.
- Maize prices were significantly higher than normal from mid-2013 until the harvest in 2014 despite an APES estimate that national production from the 2012/13 season was above consumption requirements. Price trends for commodities following harvest often do not correlate closely with production estimates.

formal trade.[3] As shown in the figure, the South African Futures Exchange (SAFEX) spot price for maize—an indicator of maize market conditions in the southern Africa region—was poorly correlated with the average monthly Malawi maize price for the period 2002 to 2014 (with a correlation coefficient

---

3 Analysis of the Trade Map dataset of the International Trade Centre on the value of Malawi's formal imports and exports for the period 2010 to 2019 shows that, on average, 23 percent of Malawi's exports by value annually went to its neighbors Mozambique, South Africa, Tanzania, Zambia, and Zimbabwe (Malawi, NSO 2020). Similar levels of exports were made to each country each year. Maize made up 1.3 percent of exports to these nieghboring countries on average annually over this period. Tea, tobacco, macadamia, sugar, and oilseeds were the largest commodity export categories.

With regard to imports over the same period, 33 percent, on average, of Malawi's formal imports by value came from these same neighboring countries annually. South Africa was the origin of 63 percent of imports from the five countries between 2010 and 2019. Maize made up 2.6 percent of these regional imports over this period. However, in most years, maize imports made up less than 1 percent of regional imports, but they spiked in years of food insecurity in Malawi, such as 2016. In terms of food imports from the region, processed foods from South Africa made up the largest category.

The unreliability of the agricultural statistics in Malawi, particularly those produced by the Ministry of Agriculture, reflect both technical capacity limitations on accuracy and operationally induced biases in the estimates.

- The APES system relies on aggregating estimates within the government's agricultural extension system from the field assistant level to the subdistrict extension planning area and up to the district and national levels. Given the relatively limited training and time that extension field assistants have to make accurate production estimates, considerable imprecision should be expected.
- The ministry is not immune to generating statistics that shed a positive light on its own performance. Notably, APES estimates of maize production immediately following the launch of the Farm Input Subsidy Program (FISP) in 2005/06 were significantly higher than those obtained from other sources (Lunduka, Ricker-Gilbert, and Fisher 2013). In a similar vein, FISP was designed to reach about half of smallholder producers. Consequently, the ministry's estimates of farm household numbers directly affect the size of the budget allocated to the program. (Chirwa and Dorward [2013] discuss in detail the problem of inconsistent or unreasonable statistics in the design and implementation of FISP.)

The deficiencies in Malawi's annual agricultural production statistics make it difficult to accurately identify deficits in food production, particularly for maize, and to plan responses to emerging food insecurity crises at a proper scale. These statistics can be reasonably used only for triangulating with data on other factors pertinent to seasonal agricultural production to estimate what actual production levels and trends might be. To ascribe more accuracy and precision to these statistics could be dangerous, particularly when using them in isolation to determine whether and at what scale to respond to food crises.

of 0.19). However, since 2015, the two price series have tracked each other more closely (correlation coefficient 0.89). Why this is the case is not clear. Restrictions on maize exports from Malawi have been in place over much of this period. Official annual imports of maize into Malawi are small—1.4 kg per person per year over the period 2010 to 2014 (Malawi, NSO 2020). Although official maize imports rose in 2016 to 17.8 kg per person, reflecting the humanitarian operations following the poor 2015/16 cropping season, 2017 imports reverted to close to longer-term levels, at 1.8 kg per person, or about 35,000 tons in total.

However, informal imports of maize into Malawi from border areas of Mozambique—much of it shipped across the border in sacks on the backracks of bicycles—are more significant. Demand for maize in heavily populated southern Malawi is relatively high, so maize produced in the border areas of Mozambique is traded informally into southern Malawi relatively efficiently (Burke and Myers 2014). There also are important marketplaces along the borders with Zambia and Tanzania that rely, in part, on flows of commodities, including maize, when cross-border price differences with central and northern Malawi provide sufficient incentive for this largely informal trade. However, because maize production in these regions is more likely to be sufficient to meet local demand than in more densely populated southern Malawi, the bulk of informally imported maize flows into southern Malawi (Whiteside 2003). The Famine Early Warning Systems Network (FEWS NET) estimated that 53,000 tons of maize annually on average came in from Mozambique informally over the period 2005/06 to 2011/12 (FEWS NET 2012). This is more than double the average annual quantity of formally imported maize reported between 2010 and 2014 (Malawi, NSO 2020). However, total annual commercial (not humanitarian) imports, including both formal and informal trade, on average constitute less than 5 percent of total maize consumption requirements in Malawi.

Household- and individual-level food security patterns

However, the reasonable levels of national maize self-sufficiency that the relatively low levels of maize imports in recent years suggest are not reflected at the disaggregated level of the Malawian household. The consumption-based poverty analysis of nationally representative data from the regular Malawi IHS includes an assessment of the ultra poor in the population, defined as those whose value of total consumption does not exceed a specific food poverty line. This ultra poverty line is the cost of a food bundle providing 2,400 kilocalories per person per day, with no nonfood consumption (Malawi, NSO and World Bank 2007b). Based on data from the IHS4 of 2016/17, 20.1 percent of the population of Malawi was estimated to be ultra poor, with total consumption valued at less than the food poverty line. There is considerable variation in the ultra poverty headcount by area: urban, 4 percent; Northern region rural, 23 percent; Central region rural, 19 percent; and Southern region rural, 29 percent. The proportion of the population whose consumption level is below the food poverty line is not changing very quickly. The IHS3 of 2010/11 showed that 24.5 percent of the total population was estimated to be ultra poor, slightly higher than the 22.4 percent found in the analysis of the IHS2 of 2004/05 (Malawi, NSO 2012).

The IHS4 also allows for categorizing households according to the severity of their recent experience with household food insecurity—61.4 percent of households in Malawi were estimated at the time of the survey in 2016/17 to have "very low food security," such that during the week prior to the interview the household experienced disrupted eating patterns and reduced food intake (Malawi, NSO 2017). The IHS4 enumeration from April 2016 to April 2017 coincided with the food crisis in the Southern and Central regions of Malawi following the poor 2015/16 harvest, so this high value should be understood in light of the widespread acute food insecurity during that period. The IHS3 of 2010/11 estimated 32.5 percent of households to have "very low food security," a value likely more reflective of chronic levels of food insecurity among Malawian households.

Both the quantity of food to which households have access and the quality of that food are food security challenges that Malawian households continue to face. Global reviews of diets show that people with lower welfare levels generally have lower dietary diversity and lower intakes of important nutrients, which adversely affect their nutritional status (Mayén et al. 2014). This is the case in Malawi. The Household Dietary Diversity Score is based on the number of food groups out of 12 from which a household reports consuming food (Swindale and Bilinsky 2006). The average score over the previous seven days for Malawian households as a whole in 2016/17, based on the IHS4, was 7.8, down slightly from 7.9 in 2010/11 (Gilbert, Benson, and Ecker 2019). Consumption of animal-sourced foods, in particular, is low—whereas about three-quarters of households reported consuming some fish, less than 40 percent consumed any meat, eggs, or dairy products. Similarly, fruit, a potentially important source of micronutrients, was reported to be consumed by only about half of all households in 2016/17.

Though not solely determined by diet quality, food security is an underlying determinant of nutritional status. The greater human development goal toward which food security is to contribute is a healthy and well-nourished population with the physical and cognitive capacity necessary to live long, creative, and productive lives. Figure 2.1 presents recent estimates of the aggregate nutritional status of young children and women of childbearing age in Malawi, along with some indications of trends. Important progress has been made in reducing the burden of malnutrition in Malawi along multiple dimensions over the past decade or so. Improved nutrition is part of what is driving important reductions in Malawi's child mortality rates—the under-five mortality rate fell from 107 deaths per 1,000 live births in 2010 to 63 deaths in 2016. More children in Malawi are growing up healthy and well nourished now than in the past. In addition to reductions in the prevalence of stunting in young children, a continuing decrease is seen in the share of women who are too thin, particularly in rural households.

TABLE 2.1 Nutritional status of young children and women in Malawi

Indicator	Malawi	North	Central	South	Rural	Urban
Stunted (low height-for-age) under-fives, 2016, %	37.1	35.1	38.2	36.6	38.9	25.0
Change 2010 to 2016, percentage points	−10.0***	−9.6***	−9.0***	−11.0***	−9.3***	−15.7***
Under-five mortality, per 1,000 live births for the 10 years preceding the survey, 2016	63	57	81	73	77	60
Change 2010 to 2016, rate	−44***	−47***	−37***	−44***	−41***	−47***
Thin women ages 15 to 49, 2016, %	7.2	5.9	6.6	8.0	7.4	6.2
Change 2010 to 2016, percentage points	−1.6**	−0.5	−1.9**	−1.6 *	−1.7***	−1.1
Overweight or obese women ages 15 to 49, 2016, %	20.7	24.1	20.4	20.2	17.2	36.2
Change 2010 to 2016, percentage points	+3.6***	+6.6***	+2.2	+3.8***	+2.9***	+8.2***

Source: Malawi Demographic and Health Survey 2010 (Malawi, NSO and ICF Macro 2011), Malawi Demographic and Health Survey 2015/16 (Malawi, NSO and ICF 2017).

Note: Asterisks show whether difference in measure between 2010 and 2016 is significantly different from zero: * = $p < 0.10$, ** = $p < 0.05$, *** = $p < 0.01$.

An emerging nutritional concern, however, is the rising numbers of adult Malawians who are overweight, as seen for women in Table 2.1. Rising overweight and obesity rates globally have contributed to increases in the prevalence of noncommunicable diseases, such as coronary heart disease, diabetes, hypertension and stroke, asthma, and some cancers. These are now among the leading causes of death in many developing regions (Shekar and Popkin 2020). The rising rate of overweight seen in Malawi replicates patterns observed in other developing countries undergoing demographic transitions, including urbanization; changes in activity levels as people are increasingly able to adopt more sedentary livelihoods; and changes in diets toward foods with more sugar, fats, and oils (Popkin, Adair, and Ng 2012; Gowshall and Taylor-Robinson 2018). Though overweight prevalence rates are rising most rapidly in Malawi's urban population, rural communities are also seeing increasing numbers of overweight and obese adults (Price et al. 2018).

Nonetheless, the trends presented in Table 2.1 support a generally hopeful outlook for nutrition in Malawi if recent progress in reducing undernutrition, in particular, can be maintained and even accelerated, and rising rates of overweight individuals are tackled. Encouragingly, reasonable progress has been made since 2010 in reducing child stunting levels—the prevalence of stunted under-fives nationally fell from 47.1 percent in 2010 to 37.1 in 2016. Differences in child stunting across districts are significant, as shown in Figure 2.3 for 2016. Likoma and Karonga districts have the lowest rates, whereas Mangochi and Neno have the highest. Of the 11 countries in eastern Africa that have conducted a Demographic and Health Survey since 2011, Malawi's stunting level falls at the median value—higher than the stunting levels estimated for Zimbabwe and Tanzania, for example, but lower than those for Zambia and Mozambique.

However, these improvements in nutrition in Malawi are not explained by better food security. As discussed in the next section, aggregate food security conditions in Malawi between 2010 and 2016 have generally been poor, particularly in the populous Southern region. Consequently, improvements in the other underlying determinants of nutritional status (Figure 2.1)—the quality of care the nutritionally vulnerable receive and improved access to public health and medical services—are more likely to have driven these nutritional advances than are changes in food security. Continued efforts are needed to ensure that all Malawians have access to enough food of high nutritional quality if further reductions in chronic undernourishment are to be achieved. Food security is central to eliminating undernutrition in Malawi.

Food Crises and Their Management

Faced with a shortage of food, particularly staples, households deplete their savings, sell assets, seek help from relatives, and change their food consumption patterns, among other coping strategies. The government also tries to help food-insecure households, particularly when there is a shock to national production. Humanitarian assistance is a regular component of the response of Malawi's government and its development partners to meet the needs of critically food-insecure households.

After each annual cropping season, the Malawi Vulnerability Assessment Committee estimates the number of Malawians likely to need food assistance through the following harvest. The estimates from 2002/03 to 2020/21

FIGURE 2.3 Stunting prevalence in under-fives by district, 2016

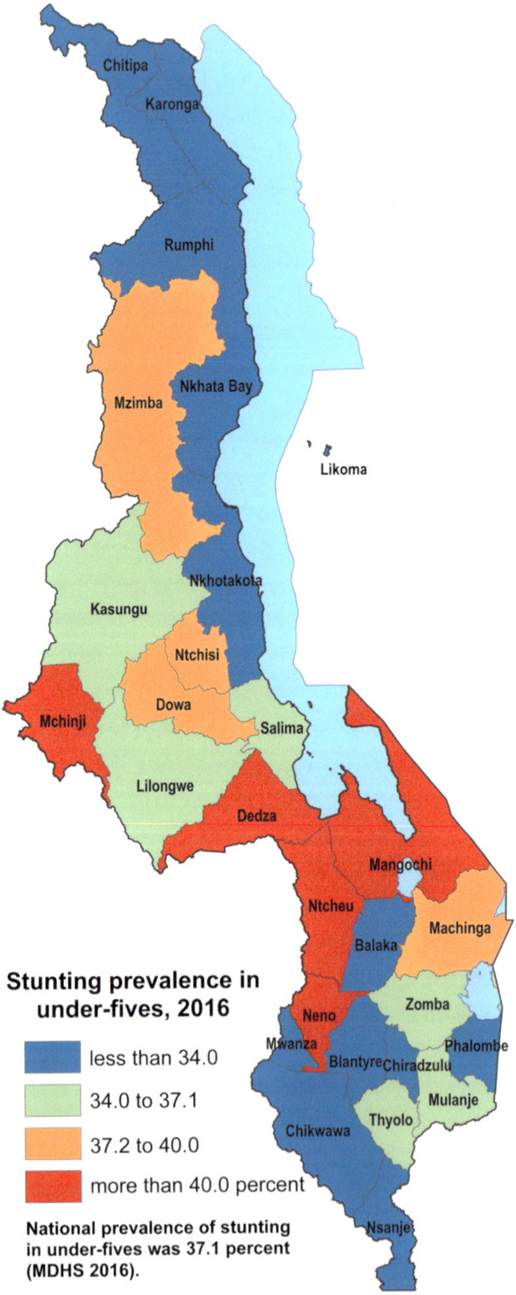

Source: Malawi Demographic and Health Survey (MDHS) 2015/16 (Malawi, NSO and ICF 2017).

are listed in Table 2.2. They are primarily based on locally disaggregated annual crop production estimates of the Ministry of Agriculture. Although the number of people in need varies from year to year, on average, 1 of 8 Malawians is judged sufficiently vulnerable to hunger to require humanitarian assistance. We see a sharp drop in the vulnerable population following the start of the Farm Input Subsidy Program (FISP) in 2005/06, a drop that was sustained for several years. In recent years, the number of people requiring food aid each year has risen, primarily reflecting both dry spells and floods, particularly in southern Malawi, which have adversely affected crop production over this period. However, this increase in the vulnerable population also reflects a reduction in the size of the input subsidy program and increasing evidence of poor implementation of the program in terms of the low amounts of maize produced for each kilogram of subsidized fertilizer distributed (Jayne et al. 2018).

TABLE 2.2 Estimates of population in Malawi vulnerable to hunger, 2002/03 to 2020/21

Year	Estimated size of vulnerable population, '000s	Vulnerable population as percentage of total population of Malawi
2002/03	3,200	28.5
2003/04	400	3.5
2004/05	1,340	11.3
2005/06	5,055	41.2
2006/07	833	6.6
2007/08	63	0.5
2008/09	613	4.6
2009/10	275	2.0
2010/11	508	3.6
2011/12	273	1.9
2012/13	1,973	13.0
2013/14	1,855	11.9
2014/15	1,312	8.2
2015/16	2,833	17.1
2016/17	6,700	39.2
2017/18	837	4.8
2018/19	3,306	18.2
2019/20	1,063	5.7
2020/21	2,642	13.7
Average	1,846	12.4

Source: Various Malawi Vulnerability Assessment Committee and Famine Early Warning Systems Network (FEWS NET) reports; National Statistical Office of Malawi population estimates and projections.

Food is the primary form of help offered to food-insecure households in a humanitarian response. The National Food Reserve Agency maintains a strategic grain reserve to cover immediate national food needs following a poor harvest while supplemental imports are organized. Although the agency has more than 280,000 tons of storage space, it generally manages reserves of around 60,000 tons of maize (FEWS NET 2018). For larger food crises, additional grain must be obtained as quickly as possible from maize markets elsewhere, particularly in southern and eastern Africa. However, obtaining grain from these regional sources can be a challenge, because national maize harvests in the region tend to move in concert due to similar rainfall patterns across the region each year. Based on data from the Food and Agriculture

Organization of the United Nations (FAO) from 1998 to 2017, the correlation coefficient for the annual national maize harvests of Malawi and those of its neighbors is 0.79 for Zambia, 0.66 for Mozambique, 0.49 for Tanzania, and 0.48 for South Africa (FAO 2020). Thus, when Malawi is looking for maize in regional markets following a poor national harvest, many of its neighbors may also be looking. Consequently, the regional market may be unable to supply enough grain for Malawi to meet its maize needs, and the country may be required to go outside of the region to obtain more maize. For example, according to the trade statistics compiled by the National Statistical Office of Malawi, for the humanitarian response mounted in 2016 following the poor 2015/16 harvest, the largest amount of maize imported came from Zambia, but Mexico was the next most important source (Malawi, NSO 2020).

Policies and Institutions around Food Security in Malawi

Conceptual frameworks in Malawi's food security–focused policy documents are not comprehensive, centering primarily on food production. For example, the government's Food Security Policy of 2006 recognizes that market mechanisms are important for ensuring food security and commits the government to establishing legislation, practices, and mechanisms to ensure competition in the market for food, food products, and agricultural inputs. But no detail then is provided regarding specific actions government will take to deepen and strengthen Malawi's food markets. The policy's principal approach to achieving national food security remains increased own production of food crops.

In terms of the institutional architecture, because food security is commonly perceived as a function of adequate food crop production, the Ministry of Agriculture is considered the center of action to ensure Malawians have reliable access to food. Reflective of this responsibility, the Ministry held the name "Ministry of Agriculture and Food Security" for many years through 2014, and in 2020 this name was restored. The political leadership of the country has consistently considered the ministry responsible for carrying out the government's duty to ensure food security for Malawi's citizens.

However, food security in the country is not synonymous with agriculture. Indeed, as will be discussed, Malawian households pursuing exclusively agricultural livelihoods are more likely to be food insecure by various measures than are households engaged in at least some nonagricultural economic activities. Within the evolving context of food security in Malawi and the demographic pressures the country faces, future policies that bear on food security will need to pay considerably more attention to how to ensure that households have access to food even if they do not have sufficient land. This requires attention to nonfarm livelihoods and occupations, the income that can be derived from such economic activities, and the strength of the markets that will be used to provide access to food for increasing numbers of predominantly nonfarm households. Malawian households will increasingly need to be able to generate sufficient cash incomes to reliably obtain the food they require from much stronger food markets. Both improving the employment choices of Malawi's workers and strengthening the markets upon which their households rely for their food are significant food security challenges facing the country.

As the importance of subsistence agricultural production for the food security of most Malawian households declines, increased use of multisectoral and multistakeholder processes will provide the broader perspective needed to guide policy reforms for food security. Expertise and perspectives from outside of agriculture can expand the channels through which a food-secure Malawi is developed and better enable coordinated strategies to achieve that vision. Food security policies, and the institutions that formulate and implement those policies, should be guided by an expanded understanding of how to ensure food security in Malawi, extending the concept beyond agricultural approaches alone.

Closely associating food security with agricultural production at a policy level also ignores many food insecurity problems in Malawi. The food security of the most vulnerable households in Malawi can be supported through specific social protection policies and actions. Similarly, to better enable households to access the food they require following significant adverse shocks to their livelihoods, the capacity of the responsible government agencies, such as the Department of Disaster Management Affairs, should be expanded. At the same time, the food security of economically productive households, whether working within or outside of agriculture, can likely best be ensured through broad economic development beyond agriculture to increase employment opportunities in the service and industry sectors of the economy. These productive households would then have a broader range of choices as to how they meet their own food needs—continuing to produce their own food, ideally at a higher level of productivity, or using their labor outside of agriculture to gain income and then relying on deep, reliable markets to obtain the food they require. Finally, trade in foodstuffs with other countries, particularly in the region, can contribute to national food security. Although it is wise for

Malawi to continue to pursue policies and implement programs to increase domestic food production, whether the country should be nationally self-sufficient in food needs to be determined on economic and risk management grounds, rather than accepted as an unexamined doctrine of national policy (Clapp 2017).

In sum, agriculture is a necessary means to household and national food security in Malawi, but neither a sufficient nor an optimal means on its own. Trade, social protection, and systems to support households in responding effectively to and recovering from adverse shocks that restrict their access to food are all required, alongside attention to food production. Moreover, based on context-specific understanding, attention will need to be paid to how particular local food systems across the country can best be strengthened using elements from this range of mechanisms to provide residents in local communities with secure access to food.

However, given the continued primacy of own production in meeting, for better or worse, the food needs of most Malawian households, and the economic importance of agriculture as a source of income, we now examine agricultural production in Malawi in more detail.

Chapter 3

AGRICULTURE IN MALAWI

Although the agriculture sector is second to the service sector in its contribution to Malawi's GDP, agriculture is central to economic and human development planning in Malawi given the significant share of the population that engages in some agricultural activities, the importance of agricultural exports for Malawi's trade, and, of relevance here, the centrality of subsistence agricultural production to the food security of most households. This chapter provides an overview of agriculture in Malawi by examining the physical basis for production and the major crops produced, particularly maize.

Several agricultural development challenges are discussed in some detail. These include the financial challenges farmers face in profitably using commercial inputs at full market cost, why food crop diversification is problematic to achieve, and the overriding constraint of low soil fertility on crop productivity levels. In considering elements of sustainable soil fertility management, the productivity levels that farmers could obtain using inorganic fertilizer versus what they do obtain are highlighted, with some discussion of what accounts for this large gap. The chapter also points out the importance of farmers' having sound information, adapted to the local context, on how best to employ soil fertility management techniques, both organic and inorganic. The discussion also covers the fertilizer subsidy programs that have been repeatedly implemented in Malawi—their justification, the general ineffectiveness of their implementation, and the opportunity costs that these fertilizer subsidy programs represent for adopting more sustainable approaches to higher crop productivity. Later sections focus on making use of Malawi's significant water resources for irrigation, customary access to arable land, and whether the distinction between smallholder farming units and somewhat larger estate farms is of much analytical or programmatic value.

The last section of the chapter provides an overview of Malawi's current policy framework for agriculture. Food security remains central to priority-setting for agricultural development, as in the past. The reverse relationship also holds—food security strategies for Malawi principally invoke agriculture-based solutions. Strategic thinking around both agricultural development and food security should seek to move the sector away from its focus on production for food self-sufficiency at both household and national levels and toward a market-centered, more concentrated and specialized sector that, nevertheless, will reliably serve to meet the food needs of increasing numbers of Malawians working outside of agriculture.

In 2019, the agriculture sector was estimated to have contributed just over one-quarter of Malawi's total GDP (World Bank 2020). Although this is a substantial drop from providing one-half of total economic output 50 years ago, globally Malawi's economy still is among the 20 national economies most dependent on agriculture. Despite significant growth in Malawi's service sector over the past 20 years, with a small manufacturing sector and limited non-agricultural natural resources to exploit economically, agriculture remains at the center of most economic production.

The food systems of Malawi historically have been based on self-sufficient, maize-focused agricultural production, and most farming is for subsistence. Nevertheless, the country's exports are also dominated by agricultural products—over the period 2015 to 2017, 87 percent of Malawi's exports by value were agricultural products, with tobacco alone accounting for 52 percent, followed by sugar and tea, with each making up about 7.5 percent of the value of all exports. Of the 10 largest export product categories, 8 are agricultural. In contrast, only 22 percent of Malawi's imports by value were agriculture-related products, and of the 10 largest import product categories, only 2 are agricultural. Of these, inorganic fertilizer made up the greatest share, at 7 percent of total imports by value. Wheat and wheat flour was the imported food product of highest total value between 2015 and 2017 (Malawi, NSO 2020). Most of the calories consumed by Malawians are from food produced in Malawi, particularly in rural communities.

Similarly, the livelihoods of most Malawians are centered on agriculture, with much of the farming done by smallholder farming households, relying on family labor to grow their own food. Employment figures disaggregated by economic sector reflect the central role of agriculture for most Malawian households. Just under 87 percent of all Malawian households and 95 percent of rural households were estimated to have been involved in some form of agricultural activities in 2013 (Malawi, NSO 2014a). The 2013 Labour Force Survey (Malawi, NSO 2014b) estimated that 64.1 percent of the employed working-age (15 to 64 years of age) population in Malawi worked in agriculture, whereas only 28.5 percent worked in the service sector. Given the share of GDP that is attributed to each sector, the relative economic productivity of Malawians engaged in farming is significantly lower than that of those engaged in the service sector. The low level of labor productivity in agriculture in Malawi reflects the development challenges facing the sector and the barriers to significant overall economic growth in Malawi emerging from agriculture.

FIGURE 3.1 Annual growth of Malawi's agriculture sector, GDP, and population, 2000 to 2019

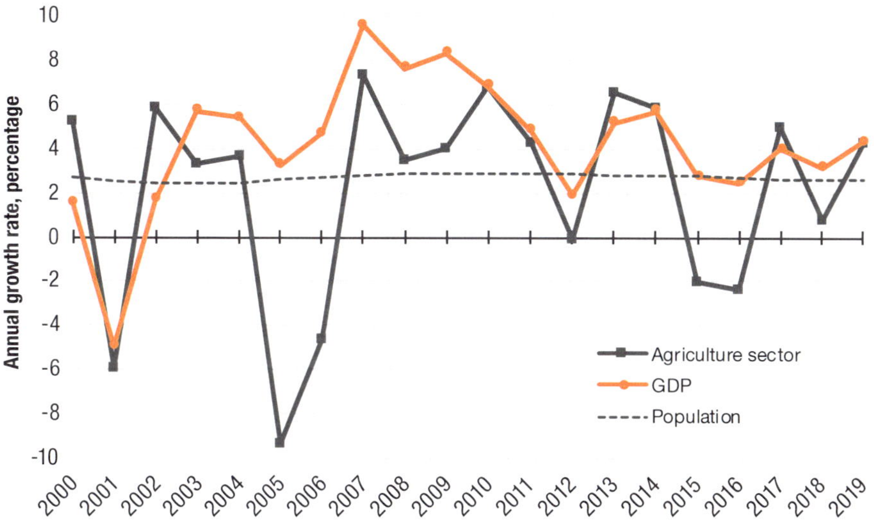

Source: Data from World Bank (2020).

Growth in Malawi's economy since 2000 has been positive overall, if somewhat erratic from year to year (Figure 3.1). Although between 2000 and 2019 the economy contracted in only one year, 2001, economic growth was slower than population growth in six years. Given the significance of agriculture in the economy, overall annual economic growth tracks the annual growth of the agriculture sector to some degree—the correlation coefficient between the two time series over this period is 0.55. Growth trends in the agriculture sector are more volatile than those of the economy as a whole, reflecting the exposure of the sector to production shocks, principally droughts or floods. Total value-added in the agriculture sector fell from year to year six times between 2000 and 2019.

But despite its importance to Malawi's economy, it is the centrality of agricultural production to the country's food system that has kept agriculture at the heart of Malawi's development strategies since the colonial era. Subsistence agricultural production remains the dominant livelihood strategy for a large majority of Malawian households. However, because the population is now growing by about 450,000 people each year (Malawi, NSO 2018), subsistence-oriented, low-input crop production on increasingly smaller household plots alone is not, on its own, a sustainable solution for feeding all Malawians. The government has repeatedly implemented large fertilizer subsidy programs for smallholders in order to raise productivity so that the national maize harvest meets the needs of the population. However, the success of these maize production programs is dependent on good rains, the burden they impose on the public budget is immense, and as will be discussed, there is no evidence that these programs result in any structural transformation of smallholder production systems or in a strengthened commercial orientation for the sector.

Similarly, liberalization of agricultural markets has been pursued in Malawi for the past 30 years—a policy reform agenda primarily pushed by Malawi's donors—in order to transform the agriculture sector away from subsistence production and better provide for the food needs of all Malawians. These efforts have expanded economic opportunities for some farmers through commercial crop production, chiefly of tobacco. However, thin and weak food crop markets, particularly for maize, coupled with recurrent weather-related production shocks, result in significant food price volatility in the country. Even though maize is produced and marketed widely throughout Malawi, the quantity of stocks that are traded commercially is low relative to overall production—estimates from analysis of the 2016/17 IHS4 for Malawi show that only about 15 percent of farm households producing maize reported selling any. These households sold on average about one-third of their maize production.

The thin maize market increases risks to household food security for the large share of smallholder farmers who rely on markets to acquire some of the maize their households consume. They cannot be confident that they will at all times be able to find in the market the maize their households require at a price they can afford. More important for the development of the sector, unreliable food markets deter specialization by smallholder farmers in commercial crop production—encouraging them to continue producing their own food and forgoing potentially significant profitable commercial opportunities—because it is unwise to depend on these markets either to buy food or to sell food. Given the high risk that they will not receive a price for the maize they produce that covers the costs of any commercial agricultural inputs used, most farming households cannot afford or forgo seeking credit to obtain those inputs. Unless they have income from other employment, households dependent on subsistence-oriented farming cannot afford the improved seed and inorganic fertilizer that could ensure larger harvests of maize and other food crops necessary to meet their own dietary needs and to generate a cash income.

The food insecurity and limited incomes common to many households in rural Malawi reflect in part the weak agricultural market system.

Physical Basis for Agriculture

Malawi is relatively well endowed with agricultural resources. Although generally depleted in crop nutrients due to continual cropping over many years, the physical structure of most of Malawi's agricultural soils is reasonably good relative to those in neighboring countries. The country receives quite reliable rainfall in the single rainy season that runs from December to April, and it possesses significant water resources, most notably Lake Malawi and the Shire River that flows out of the lake, which could form the basis for extensive irrigated farming zones. Precolonial settlement patterns and the spatial variability in current rural population densities across southern Africa reflect the relatively advantageous agricultural resources found in Malawi. In part because of the agricultural potential of the country, the population of Malawi has increased significantly over the past century—from an estimated 970,000 inhabitants in 1911, to 4.0 million in 1966, to 17.6 million at the last census, that of 2018 (Malawi, NSO 2018). In consequence, farms have shrunk. As estimated based on IHS4 data, the mean land area now cultivated by farming households in Malawi is relatively small in the context of Africa south of the Sahara, at 0.65 ha per household that reported growing any crops.

TABLE 3.1 Rural population densities, cropped land as share of agricultural land, and livestock production as share of agricultural production, Malawi and neighboring countries, 2016

Country	All land area, persons per sq. km	Agricultural land area, persons per sq. km	Arable land and land under permanent crops as share of all agricultural land, %	Livestock production as share of gross production value of agriculture, %
Malawi	158	257	68	15
Mozambique	24	38	12	20
South Africa	15	19	13	51
Tanzania	41	92	39	18
Zambia	13	39	16	43
Zimbabwe	27	65	25	42

Source: Data from FAOSTAT (FAO 2020).
Note: The population densities are based on the total rural population of each country and not on the actual number of residents on agricultural land. Agricultural land area is the sum of the area in the country classified as arable land (land under temporary agricultural crops or temporarily fallow), the area under permanent crops, and the area under permanent meadows and pastures.

Rural population densities in Malawi are significantly higher than those of its neighbors, which have considerably larger portions of their land area that are unsuited for cropping (Table 3.1). A much larger portion of the agricultural land in Malawi is under temporary and permanent crops than in neighboring countries, where most of the agricultural land is under permanent pasture and is not used for crops. Reflecting this difference, the share of livestock in Malawi's total agricultural production is considerably smaller than it is for other countries in the region.

Rainfed crop agriculture is practiced throughout the country. The most productive areas are found in the mid-altitude plateau zone above the escarpment of the East African Rift Valley that runs the length of the country. This zone generally receives between 800 and 1,000 mm of rainfall annually. However, even in the somewhat less productive or riskier agroecologies of the lakeshore and in the Lower Shire Valley in the far south of the country, crop agriculture is the predominant economic activity. Figure 3.2 presents crop suitability maps for shorter-duration maize varieties and for air- and fire-cured tobacco grown under improved smallholder production practices.[1] Except in the rugged terrain and thin soils of the Rift Valley escarpment areas, the Middle Shire Valley, and the Mulanje and Nyika highlands, as well as the marshy areas around Lake Chilwa and in the Lower Shire Valley, virtually all the rest of the country has soil and climate characteristics that are at least marginally suitable for the production of these two important crops, with most areas judged to be moderately suitable.

However, agriculture in Malawi is vulnerable to production shocks, particularly extreme weather events, including droughts and floods. Annual losses from shocks to production from 1980 to 2012 were estimated to average $149 million (Giertz et al. 2015). The Malawian economy in aggregate also falters under such conditions, in which commodity production is reduced and

1 These crop suitability maps were developed by the Land Resources Evaluation Project, which was implemented from 1988 to 1992 as a joint effort of the government of Malawi, the United Nations Development Programme, and the Food and Agriculture Organization of the United Nations. The project involved a close reconnaissance of the agroecological resources of Malawi. Extensive fieldwork was done to map the soils across the country at the relatively detailed scale of 1:250,000 (1 cm = 2.5 km). Weather data were analyzed to generate averages for various indicators for use in mapping the agroclimatological zones of Malawi at the same geographic scale. The soils and agroclimate maps were then overlaid to develop a "land unit" map for Malawi. Each land unit was defined by a unique combination of relatively homogeneous soil and climate properties within its boundaries. These land units were then used with information on the optimal soil and climate conditions for growing a range of rainfed agricultural crops, as well as irrigated rice and tree species, to conduct a spatial suitability analysis for the production of each crop or tree species in each land unit. For additional detail on the Land Resources Evaluation Project and the outputs it generated, see Benson, Mabiso, and Nankhuni (2016).

FIGURE 3.2 Crop suitability maps for Malawi for short-cycle maize and air- and fire-cured tobacco under improved traditional management practices

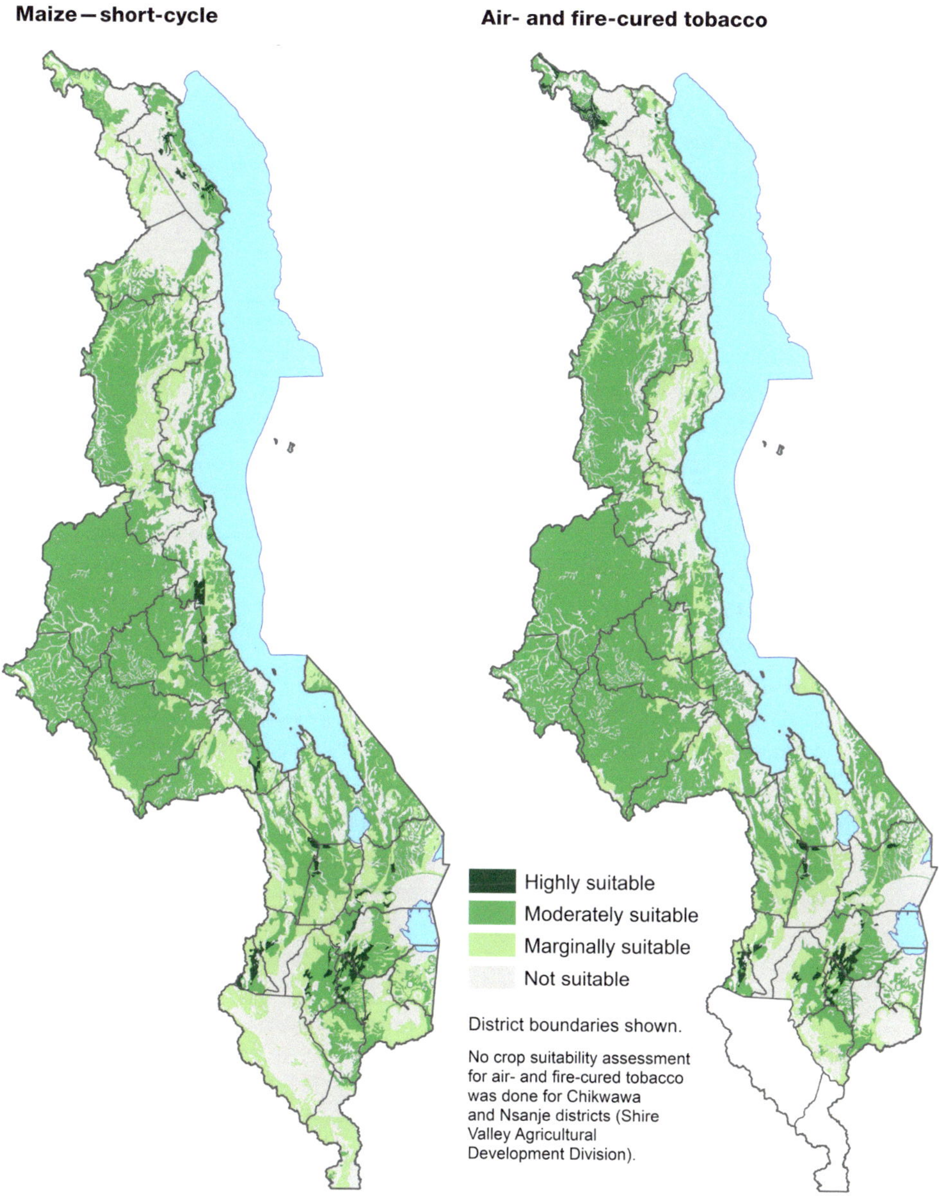

Source: Benson, Mabiso, and Nankhuni (2016). Analysis and mapping of information from Malawi Land Resources Evaluation Project, 1988-1992 (Eschweiler et al. 1991).

resources, whether public or private, that might have been used for productive investment are diverted to address the immediate crisis of food insecurity facing many Malawian households. Many of the economic gains made by farm households and by the Malawian economy over years of normal to good crop production are swept away in the aftermath of a poor harvest.

It is estimated that global climate change will result in increases of between 1.0°C and 3.0°C in monthly mean maximum daily temperatures across Malawi between 2000 and 2050 (Saka et al. 2013). These increased temperatures will adversely affect the production of crops that are not heat tolerant. However, given the varied topography of the country and the moderating effects of Lake Malawi on local climates, estimates across climate change models of how precipitation patterns might change in Malawi are not consistent. Although variability in rainfall both within and between rainy seasons is expected to increase (Grist 2015), there still is considerable uncertainty as to just what global climate change will mean for Malawi in terms of agricultural production and climate-induced shocks. Most observers are pessimistic.

With regard to other risks to agriculture, most farmers in Malawi are not significantly exposed to market-induced economic shocks caused by global commodity price movements. This reflects the limited involvement of these farmers in international agricultural markets rather than any inherent resilience in commercial agriculture in Malawi. Given the manner in which their value chains are organized and the significant transaction costs, particularly for transport, that they face in moving their produce to international markets, Malawian producers tend to receive low prices relative to international prices regardless of international market conditions. In addition, the marketing systems in Malawi through which the cash crops that smallholders are likely to produce are aggregated for export are not competitive, with only a few exporters operating. This is particularly the case for burley tobacco (Prowse and Moyer-Lee 2014). Producers have limited bargaining power in the face of low price offers for their crops.

Agricultural Production

The agriculture sector in Malawi can be disaggregated by the objective of production—between food production, including livestock, on the one hand, and cash crop production on the other.

Food production

MAIZE

Food production is central to the livelihoods of most rural households in Malawi. Maize is the principal food crop produced and the dominant staple food in almost all districts of the country. The crop was already incorporated into Malawian farming systems by the late 1800s, having been introduced from its center of origin in Mexico to coastal southern Africa, likely by the Portuguese, and then disseminated inland. At the start of the colonial era, maize was cultivated alongside other grains, particularly sorghum and finger millet. However, maize progressively came to dominate smallholder farming and food systems in Malawi (Vail 1983) beginning during the colonial period and continuing following independence in 1964. Malawi's current annual maize consumption is estimated at about 2.75 million tons (Babu et al. 2018). Other staples are much less commonly consumed. Today, only in Likoma district (which comprises two islands in Lake Malawi) and in areas along the central lakeshore in Nkhotakota, Nkhata Bay, and Rumphi districts does one find cassava as a common staple, generally alongside maize. Also, in the Lower Shire Valley, sorghum remains an important staple together with maize, being well adapted for the growing conditions there.

"White flint" is the preferred maize type in Malawian diets. Starting in the 1950s, agricultural researchers worked to develop improved varieties of maize adapted for Malawi, including hybrids. Although the early hybrids were softer dent varieties, by the late 1980s semi-flint varieties that better met the consumption preferences of the population had been released to farmers. More recent maize breeding efforts in Malawi have focused on disease and pest resistance, and on water use efficiency for drought tolerance. Propelled by maize technology extension programs and various subsidy programs, uptake of improved maize varieties has been reasonably good—the 2014 Welfare Monitoring Survey found that 68 percent of all households producing maize grew at least some hybrid varieties (Malawi, NSO 2015).[2] Several international seed companies operate in Malawi, with hybrid maize central to their business (Abate et al. 2017).

2 Although F1 hybrid maize seed, in order for the crop to maintain its hybrid vigor, should be purchased new each year and not recycled, many farmers in Malawi recycle some of their harvested maize (F2 and subsequent generations) as seed in the following season, even if the maize was produced from F1 hybrid seed (Morris, Risopoulos, and Beck 1999). However, in surveys, farmers often will report such recycled seed as "hybrid," even when technically it is not (Wineman et al. 2019). Consequently, survey estimates of hybrid maize seed use by smallholder farmers in Malawi, as in much of Africa, overstate somewhat the use of F1 hybrid seed.

In parallel with maize breeding efforts, maize agronomy research has focused on soil fertility management for sustained higher levels of maize productivity. Maize production has been shown to be very responsive to the application of nitrogen and sulfur on most soils in Malawi, as well as to phosphorus, if less reliably so (Benson 1999). Such yield gains from the application of inorganic fertilizers require that the fertilizer be combined with good crop management practices that include use of improved seed adapted for prevailing agroecological conditions and resistant to common maize diseases and pests; timely planting, fertilizer application, and weeding (at least twice); somewhat higher plant populations than traditionally used; and, to maintain soil health through adequate soil organic carbon levels, the use of high-quality (low carbon-to-nitrogen ratio) organic materials, such as crop residues from grain legumes grown in rotation with maize, animal manure or compost, and biomass from agroforestry species (Malawi, MoAFS 2012; Snapp et al. 2014). The significant yield gains possible through the application of inorganic fertilizer were demonstrated to smallholder farmers in the 1980s and 1990s. However, as will be discussed, many smallholders are unable to manage their fertilized maize crop in an optimal manner due to their inability to afford the full amount of purchased inputs they require, late input acquisition, insufficient labor to adequately weed and otherwise manage the maize crop in the field, and for many, poor access to information on how best to use what fertilizer they do obtain.

Given the high price of fertilizer relative to that of maize, maize price volatility, and the position of most farmers as price-takers in agricultural output markets in Malawi, the production of fertilized maize even under good management has not been a reliable commercial enterprise. Although most Malawian farmers seek to use fertilizer in their maize production, most are unable to afford it, due to cash constraints and very limited access to credit. Only those with sufficient cash flow from other productive activities, particularly in the formal wage sector, are able to afford fertilizer. Moreover, those who use income from other economic activities to buy fertilizer generally do so with the objective of producing maize for their own consumption, rather than for sale. Under most cropping and market conditions, this makes financial sense—these households are implicitly valuing the maize harvest that they will consume at the higher consumer maize price (what they would have to pay in the market for this maize in the period of high prices before the following maize harvest), rather than the significantly lower producer price (what they would receive if they were to produce it for sale). For the Malawian smallholder under current conditions of thin maize markets, growing fertilized maize for own consumption is cheaper than relying on the market to meet household maize needs (Benson 1999). At the same time, growing fertilized maize for commercial sale without the benefit of subsidies or other distortions to agricultural input or output markets generally will not be the optimal cropping choice for the profit-seeking farmer in Malawi and, indeed, will often result in financial losses. This point is discussed in more detail later.

Most smallholder farmers, excepting those with income from other crops or nonfarm employment, do not have sufficient financial resources to pay full market price for the fertilizer they require for their maize plots. Without adequate access to fertilizer, households that rely on their own maize production

TABLE 3.2 Selected annual food crop production in Malawi, 2013/14 to 2017/18, averages for area planted, production, and yield, plus potential maximum yields

Crop	Area planted, '000 ha	Total production, '000 mt	Yield, kg/ha	Potential maximum pure-stand yield under smallholder conditions, kg/ha
Maize, all	1,693	3,057	1,800	7,000 (hybrid)
Irrigated	*172*	*463*	*2,680*	n.a.
Estate production	*44*	*125*	*2,830*	n.a.
Rice	63	112	1,770	5,000 (irrigated)
Sorghum	383	340	890	1,700
Millet, pearl and finger	99	81	820	2,000 (both species)
Groundnut	52	32	620	2,500
Soyabean	156	155	990	4,000
Common bean	331	186	560	2,000
Pigeonpea	245	386	1,570	2,500
Cowpea	88	39	440	2,000
Other beans or peas	35	30	860	Various
Cassava	212	n.a.	5,130	30,000 (fresh)
Sweet potato	211	n.a.	2,600	30,000 (fresh)
Irish potato	62	n.a.	3,500	20,000 (fresh)

Source: Area, production, and yield are from the annual agricultural production estimates of the Ministry of Agriculture for total national production (smallholder rainfed and irrigated, plus estate), except for yield figures for cassava, sweet potato, and Irish potato, which are from Malawi, NSO (2010); potential yields are from Malawi, MoAFS (2012).

Note: The estimates from the Ministry of Agriculture for total production and yield for cassava, sweet potato, and Irish potato are judged not plausible given the low level of consumption of these crops that has repeatedly been recorded in household consumption surveys in Malawi (see Box 2.1). The estimates of potential maximum yield assume optimal crop management, use of improved planting materials, and application of sufficient crop nutrients to achieve the production levels specified. mt = metric tons; n.a. = not available.

to feed themselves are at significant risk of running short of maize during the lean season before the following harvest. Consequently, over the past 30 years, the government of Malawi has repeatedly provided significantly subsidized fertilizer to smallholder maize producers.

Most Malawian farmers achieve maize productivity levels that are well below the potential yields (Table 3.2). Maize hybrids grown in the mid-altitude plateau agroecology, where much of Malawi's maize is grown, have the genetic potential to yield up to 7.0 tons/ha with appropriate input use and good crop management. However, given the extensive nutrient depletion in most soils cropped by Malawian smallholders, it will take many years of good soil husbandry to bring those soils back to a condition in which such potential yields can be approached. Most smallholders will be unable to make the long-term investments needed to restore the full productive qualities of the soils they farm, particularly if these investments entail significant opportunity costs in terms of meeting their immediate food needs. Consequently, somewhat lower potential maize yields, of 4.0 to 5.0 tons/ha, are a more reasonable maize productivity target in the longer term. Nonetheless, with the provision of well-informed, locally adapted, and widely available agricultural extension messages; continued breeding efforts to improve the maize varieties available to farmers; continued research on proper crop management; timely and reliable access to inputs through efficient markets; and sufficient and reasonably predictable price incentives in output markets, maize producers in Malawi can achieve significant increases over present yields, contributing to feeding the country with its staple food for some time to come. Continued investments in raising farmers' productivity, for both maize and other crops, should remain a central element in any effort to develop Malawi's agriculture sector and to increase the amount of maize and other food crops available in domestic markets.

OTHER FOODS

Although maize dominates the diets of most Malawian households, a range of other food crops is produced and consumed. Breeding to develop improved, more adapted planting materials has been done for most of these crops, but rarely is fertilizer applied to them. Other cereals besides maize include rice, which is grown primarily under lowland conditions along the lakeshore; sorghum; finger millet; and some pearl millet. There is potential for wheat production, particularly in highland areas, but only small amounts are grown. Consumption of rice and wheat is increasing among urban households. A small portion of the rice is imported, whereas all the wheat comes from outside of Malawi.

Other significant staple food crops include roots and tubers. For the last 15 years, production estimates by the Ministry of Agriculture suggest that by weight, somewhat more cassava and sweet potato than maize is produced annually in Malawi. However, household consumption surveys show that less than 5 percent of the calories consumed by the average Malawian come from either cassava or sweet potato (Gilbert, Benson, and Ecker 2019). This significant difference between production and consumption estimates reflects both the challenges of collecting accurate production statistics for roots and tubers (see Box 2.1) and the practice among many producers of cassava of planting it as a reserve crop that is harvested and consumed only if household maize stocks are depleted (Haggblade and Zulu 2003).[3] Sweet potato is a traditional crop grown in seasonal wetlands. The orange-fleshed types have been extensively promoted in recent years to increase vitamin A consumption. Irish potato is commonly grown in highland areas with reasonably good transport connections to urban centers, thus serving as both a food crop and a cash crop for producers.

Grain legume crops, or pulses, serve as both cash and food crops. Their production has received considerable attention from agricultural researchers in Malawi seeking to develop maize–grain-legume intercropping or rotational farming systems as an element of sustainable soil fertility management, although so far without significant adoption by farmers. Most groundnut is consumed domestically, but the crop has experienced booms and busts in production, primarily linked to changing demand from regional and international export markets. Similarly, soyabean and pigeonpea are of interest primarily as cash crops. Several different bean and pea types are grown for consumption, with the most prevalent being common bean (*Phaseolus vulgaris*). Typically, both the grain and the leaves of these crops are used in relishes to eat with *nsima*, the thick maize meal porridge that is the principal form in which maize is consumed in Malawi.

---

3 Estimating agricultural statistics for cassava, in particular, is challenging (Keita 2003). Its cropping cycle can go beyond a year. It is frequently planted as an intercrop, complicating estimation of planted area. Cassava roots do not store well without processing, so they cannot be held for later measurement of harvest. Whether a wet or a dry root weight is used can result in significantly different estimates of production. In some food systems, cassava leaves may be more important for diets than are the roots. These factors all reduce confidence in agricultural statistics on cassava production and also account for significant differences between sources, such as between national crop production estimates and those from household surveys.

Although making similar estimates for sweet potato and Irish potato are not as difficult as for cassava, many of these same challenges arise.

Vegetables and fruits are produced in household gardens for home consumption and in peri-urban areas to serve urban markets. These foods are quite important for improving the quality of local diets, being potentially important sources of micronutrients, although much greater consumption is warranted. Of the vegetables, green leafy vegetables are the most commonly consumed and also the most significant nutritionally. The value chains associated with vegetables and fruits are rudimentary, although a commercial firm has begun exports of fruit purees and juices in the past five years.

Compared with its role in neighboring countries, livestock in Malawi is not an important component of the agricultural economy. Reflective of this, on average Malawians consume relatively small amounts of nutrient-dense animal-sourced foods—an estimated 13.4 percent of all protein consumed in 2016/17 came from animal-sourced foods. Consumption of fish is increasing, whereas that of meat and poultry is declining—fish became a more important source of protein than meat for the average Malawi household between 2010/11 (IHS3) and 2016/17 (IHS4) (Gilbert, Benson, and Ecker 2019). Though still low overall, urban consumption of animal-sourced foods on a per capita basis is about double that in rural households. The limited livestock and livestock products are primarily produced for the domestic market—few are exported.

In terms of livestock produced, poultry in particular can be an important secondary element in the agricultural livelihoods of smallholder households. Larger animals are more likely to be part of the farming systems of wealthier households in areas with adequate grazing land, as in the Northern region and in the Lower Shire Valley in the far south. Expansion of dairy production in peri-urban areas around the cities of Blantyre, Lilongwe, Zomba, and Mzuzu, particularly among smallholders with one or two cows, has been a focus of government and donor efforts over the past 15 years. In 2008, it was estimated that nationally there were 30,000 dairy cattle owned by about 8,000 farmers (Chagunda et al. 2010). Livestock markets in Malawi appear to be improving, possibly driven by increasing demand from urban consumers. With local adaptation in terms of animal mixes and scale of production to reflect locally available resources, livestock husbandry has unrealized potential for contributing to increased and more diverse commercial agricultural production (FEWS NET 2016).

FOOD CROP DIVERSIFICATION

Over the years, regular calls have been made to diversify the food system of Malawi away from its heavy reliance on maize. Several pilot efforts have been justified based on the need to do so. Diversification of the country's food production is expected to be driven primarily by increased consumption of grain legumes, vegetables, and fruits, as well as animal-sourced foods. If successful, diversification efforts would provide nutritional benefits, with greater individual consumption of fats, protein, and a range of micronutrients, and increase the resilience of Malawi's food systems, particularly to shocks to maize production, but also to periods of volatility in prices.

However, diversifying food production away from maize is not an easy task given the broad set of advantages of maize in the Malawian context and the primary nutritional demand for the carbohydrates it provides. Agroecologically, the maize varieties grown in Malawi are well adapted. Erratic rainfall patterns historically have significantly affected national production once or twice a decade, but with improved maize seed, sufficient nitrogen applied to maize plots, and good crop management, farmers should be able to consistently produce more than 3.0 tons/ha in most years, with even higher yields possible. Maize has strong cultural significance in most communities across the country—although several available carbohydrate-dense crops could substitute for maize, culturally those crops are considered imperfect substitutes. The flint maize types preferred in Malawi store well over the long annual dry season if the harvested grain is kept dry and ventilated. As a cereal, maize is reasonably easy to transport and to market without losses in quality. Although primarily grown for own consumption, maize will always find a market and can be considered a semi-tradable crop with potential for export to neighboring countries. Sales of maize are among the strategies farming households may consider for meeting their cash needs. Though this is the case with all foods produced by smallholders, there are more limited marketing opportunities for cassava, for example, which is an obvious substitute for maize in meeting the carbohydrate needs of Malawians.

Given the perceived advantages of maize production and consumption and in a context of limited land availability, a farmer's decision to plant some of their land in crops other than maize is not taken lightly. In the absence of locally adapted information for both producers and consumers on the benefits of alternative food crops; with weak markets that constrain commercial production of maize also constraining such production of other food crops; and with continuing low maize productivity levels leaving relatively little cropland available for the production of other crops, maize will continue to dominate most smallholder fields, food systems, food policies, and political discourse in Malawi.

Export crops

Although maize production dominates the agricultural landscape in Malawi and food security concerns drive much of the agricultural development planning efforts in the country, export crops also play a significant role in the overall economy. In 2004, export crops were estimated to account for just over one-quarter of Malawi's agricultural GDP and just over 10 percent of total GDP (Benin et al. 2008). Their contribution to Malawi's export earnings is considerably more significant, accounting for 87 percent of Malawi's exports by value between 2015 and 2017 (Malawi, NSO 2020).

TOBACCO

In both 2016 and 2017, about 150,000 tons of tobacco were exported with a value each year of about $540 million. Minimally processed burley and flue-cured tobacco leaf accounted for just over 52 percent of the value of all exports between 2015 and 2017. Tobacco has been the principal cash crop produced in Malawi since quite early in the colonial period, with commercial production starting in the Shire Highlands in the south of the country in the 1890s. It now is produced in most areas of the mid-altitude plateau in all three regions of Malawi, with most intensive production in Lilongwe, Kasungu, and Mchinji districts in the Central region.

After the introduction of tobacco as a cash crop by colonial European planters on their estates, smallholders became involved in its production quite quickly, initially as seasonal tenant farmers on estates. Tobacco production regulations in both the colonial and early independence periods in Malawi privileged estate production of tobacco. But when these regulations were modified to allow increased involvement of smallholders in tobacco production, particularly of fire-cured dark leaf initially and, now, air-cured burley leaf, both of which require little capital investment, the supply response of smallholders was significant (Green 2012b). Thereafter, the historical pattern has been that, as smallholder production began to dominate the tobacco sub-sector, the state would reimpose regulations advantageous to estate production, and production would revert to the estates. Prowse (2013) documents several cycles of this pattern over the past 100 years.

Malawi is in the middle of one such smallholder-dominated tobacco production period now, but this one, which began following the liberalization of tobacco production in the mid-1990s, might be permanent. The sharp uptake in burley tobacco production by smallholders since about 1995 has led to a crisis in the tobacco estate sector, which is no longer able to competitively produce burley in part because estates are unable to obtain sufficient tenant labor to do so. Estates now primarily produce only the more capital-intensive flue-cured tobacco. The annual crop estimates for the period 2013/14 to 2017/18 show that estates produced about 11 percent of the quantity of tobacco produced nationally (but about 48 percent of all flue-cured tobacco). Twenty-five years earlier, in 1990, the same production data series showed that smallholders were responsible for only 13 percent of total tobacco production.

Until quite recently, all tobacco produced in Malawi legally had to be sold bale by bale in a competitive auction system at a handful of auction floors established in tobacco-producing areas of the country, the largest in Lilongwe. However, there have never been more than 10 tobacco firms buying from the country's auction floors, with one or two of them accounting for most tobacco purchases in any year. Consequently, a fundamental concern about the tobacco output market in Malawi has been its oligopsonistic structure and evidence of collusion among buyers on pricing the leaf to the detriment of producers (Otañez, Mamudu, and Glantz 2007; Prowse and Moyer-Lee 2014). In consequence, prices that Malawian smallholder producers receive for their leaf do not closely reflect international tobacco prices.

Since about 2000, the government has increasingly allowed tobacco-buying firms to enter into direct production contracts with smallholder producers—an arrangement called the Integrated Production System (IPS)—rather than being required to purchase their leaf only on the auction floors. Under the IPS, tobacco firms provide inputs and agronomic advice to producers and maintain a high level of control over the production process during the growing season (Moyer-Lee and Prowse 2015). Tobacco firms are motivated to use this system by increasing demands from cigarette manufacturers for traceability of the leaf the firms supply to the manufacturers. IPS sales now dominate in Malawi's tobacco industry—82 percent of the burley crop and 91 percent of the flue-cured crop were sold under the IPS system in 2015. Average prices for burley were 17 percent higher when sold under IPS than when sold at auction, although there is evidence that the tobacco presented to the auction floors was of significantly lower quality (Chirambo 2016). Although smallholder tobacco producers can reliably obtain inputs and advisory services under the IPS contracts, their bargaining power in the marketplace is even more restricted than on the auction floors, because they deal with only one firm.

The proper regulatory role for government in providing oversight of IPS and other contractual relationships within farming of all sorts is an area of ongoing policy debate in Malawi. The Ministry of Agriculture developed a National Contract Farming Strategy in 2016, following several years of discussions (Malawi, MoAIWD 2016a). This strategy proposed a legal framework to better manage contracts and mitigate the risk of noncompliance by either party—the contracted farmer or the buyer—in order to create mutual benefits. It proposed mechanisms involving third-party oversight of dispute resolution between the contracted parties. However, the strategy has not advanced to become a policy upon which a legislative framework could be established. This in part reflects a pessimistic view of contract farming held by many stakeholders in the agriculture sector, particularly their distrust of the international tobacco leaf merchants who have promoted contractual arrangements with smallholder farmers in Malawi very successfully over the past 20 years (Prowse and Grassin 2020). Concerns remain about the organization and costs of fairly implementing the dispute resolution procedures within any legislative framework governing contract farming arrangements.

OTHER EXPORT CROPS

After tobacco, tea and sugar are Malawi's most important exports. Both are primarily produced on estate farms. Tea has been produced for more than a century on three to four dozen estates in high-rainfall areas of Mulanje and Thyolo districts, plus several estates in Nkhata Bay. Most of the tea estates are owned by international agribusiness firms. Sugar is produced on two irrigated, vertically integrated estates owned by the international sugar producer Illovo—one in the Lower Shire Valley in Chikwawa district at Nchalo and the other at Dwangwa on the lakeshore in Nkhotakota district. A third sugar estate began production in 2017 in Salima district. The sugar estates and a few of the tea estates obtain a small part of their cane and green tea from smallholder outgrowers in the vicinity, purchasing both commodities on the basis of seasonally negotiated pricing formulas.

Cotton is produced by smallholders along the lakeshore and in the Upper and Lower Shire Valleys. These producers sell to about a dozen ginners. Cotton has been long been promoted as an export crop for Malawi, receiving considerable government attention beginning early in the colonial period and continuing since independence. During the Kamuzu Banda era, a parastatal textile factory, known as David Whitehead, operated in Blantyre and absorbed a significant share of domestic cotton production. However, the firm closed in the mid-1990s following economic reforms that resulted in increased competition from imported textiles and clothing, particularly secondhand clothing, and has not been able to sustainably restart production since. Cotton prices have been variable over the years, with consequent booms and busts in production. More than with the other export crops, international cotton price swings are transmitted to local producers, because the ginners sell their generally minimally processed product to the international market at those prices. These international price movements, when coupled with variability in production due to weather and pests, result in significant instability in prices, and hence production volumes, from year to year.

The National Export Strategy, formulated in 2012 and revised in 2020, designates oilseeds—particularly soyabean, groundnut, and cottonseed—as one of its three product clusters for generating significantly increased export revenues for the country (Malawi, MoIT 2012, 2020). Of these crops, groundnut has the longest history in Malawian farming systems, produced both for processing into cooking oil domestically and regionally, and for overseas confectionery markets. In the 1990s and early 2000s as aflatoxin contamination regulations limited access to some of Malawi's traditional groundnut export markets, groundnut exports to overseas confectionary markets fell sharply. Efforts have been made since then to address the problem of aflatoxin contamination in groundnut production and processing in Malawi, with some of these efforts supported by investors from the export markets. Malawi has begun reentering those overseas markets. However, there also has been a shift in groundnut export patterns over the past 10 to 15 years, with most exports of Malawi groundnut now going to regional markets in southern and central Africa (Edelman and Aberman 2015).

Soyabean is produced by both smallholders and estates. Subsistence use of soyabean is almost nonexistent. As a cash crop, there is growing demand both domestically and regionally for soyabean from livestock producers for feed, particularly for poultry, and from food producers for cooking oil and other soya products. Over the past decade, the government has occasionally imposed export licensing restrictions on soyabean to safeguard access to local soya at low cost for the domestic poultry industry and cooking oil processors. Because these interventions have constrained exports, the average annual value of soyabean exports from Malawi has been variable. Although soyabean exports were valued at over $19 million in 2017, in the two years prior, their export value did not exceed $9 million per year (Malawi, NSO 2020).

More minor export crops include pigeonpea, which is grown by smallholders primarily in southern Malawi and is sold into the South Asia market through Blantyre-based processors, and macadamia nuts, grown primarily on tea estates as a secondary crop. Coffee also is exported from a few estates in the south and several thousand smallholder producers in the Northern region, but the subsector is not growing significantly.

Fertilizer

Sustainable soil fertility management is a long-standing and intensifying challenge to improving crop production in Malawi. For subsistence-oriented smallholder farming households, the declining fertility of their cropland also increasingly undermines their food security. Considerable research has been done over more than 50 years to identify maize-legume crop rotation, intensive intercropping, green manure, or agroforestry-based cropping systems that are effective in maintaining adequate plant nutrient levels in the soil; do not exacerbate the labor, land, or financial constraints that smallholder farmers in Malawi face; are commercially viable; and consistently meet the dietary needs of these households. Although important insights have been gained on how such primarily organic approaches to soil fertility management perform, significant uptake of such alternatives to the dominant maize cropping system (with some intercropping) has not occurred. None has consistently proved superior to the use of inorganic fertilizers across the performance metrics noted.

In consequence, inorganic fertilizer is viewed by most farming households in Malawi as critical to realizing improved livelihoods from their farming activities and to assuring their own food security. Moreover, ensuring that all smallholder farmers have access to fertilizer consistently figures in the election platforms of political candidates in Malawi. For the country's political leaders, access to inorganic fertilizer for Malawi's farming households has been one of the most important issues over the past 30 to 40 years, and for the past 15 years, a significant share of the annual government budget has gone to subsidizing fertilizer for smallholder farmers.

Crop response to fertilizer

In the precolonial and early colonial periods, smallholder agricultural systems in most areas of Malawi relied on shifting cultivation or crop-fallow systems for sustainability. Land was opened for production for several years, and when crop yields declined, new uncropped land in the vicinity was brought into production. The depleted land was left in fallow for many years, allowing crop nutrient stocks to rebuild (Allan 1965). However, the population of Malawi has increased significantly over the past century. In consequence, this land-extensive method for sustaining crop productivity is no longer practiced to any significant extent—there is simply not enough land in Malawi to do so. Households generally farm the land to which they have use rights every year, using family labor and low levels of other inputs. Given the strong subsistence orientation of most producers, the farming landscape in Malawi is dominated by maize, which is consumed in the household.

As a result, farmers remain cash-poor and unable to purchase inorganic fertilizer to improve their crop yields. Over the generations, most of the plant nutrients in the soils, particularly nitrogen, have been mined out through the growing of repeated maize crops, resulting in low crop productivity levels of around 1.0 tons/ha for unimproved local maize varieties grown without fertilizer.

Few farmers practice crop rotations that alternate maize with grain legumes—principally groundnut, pigeonpea, and soyabean—crops which fix atmospheric nitrogen and make it available through the crop residue to subsequent crops on the same plot. Although such rotations sustain soil fertility better than continuously monocropped maize, output markets for both maize and grain legumes remain too uncertain for most farmers. Because farmers cannot be confident that they will be able to purchase the maize that their households require during the grain legume period of such a crop rotation with the proceeds from the sale of their grain legume harvest, many choose to avoid that market risk and devote their land to continually producing the maize their household needs (Dorward et al. 2009). Another common method used globally to manage crop nutrient levels in agricultural soils, the manuring of fields, is similarly problematic in the Malawi context given the relatively small numbers of livestock. Most of the land that could be used for pasture is already being used for crops. With limited use of crop rotations by farmers and few other organic sources of plant nutrients, such as manure or agroforestry species, nitrogen, the soil nutrient most needed by maize, has become the principal constraint to increased maize production. Urea fertilizer, particularly, is an effective solution to this constraint.

This view of inorganic fertilizer as a central tool for raising agricultural productivity in Malawi has a relatively long history. The colonial government provided farmers with subsidized fertilizer starting in 1952 to improve the quality and yields of tobacco and to raise maize productivity so that land could be freed for legumes in rotation (Kettlewell 1965). Efforts to increase the uptake of fertilizer, including fertilizer subsidies, were part of the policy of the government of Kamuzu Banda, both to accelerate estate-led production of cash crops and for broader food security objectives. Fertilizer has remained a central policy concern for all subsequent governments.

Extensive research programs on fertilizer response in maize were conducted in the late 1950s and early 1960s (Brown 1966) and again in the mid-1970s (Bolton and Bennet 1975). The most recent national effort evaluating maize yield response to fertilizer was done in the 1995/96 and the 1997/98 cropping seasons under the Maize Productivity Task Force of the Ministry of Agriculture

(Benson 1999). Researchers from the Chitedze Agricultural Research Station outside of Lilongwe worked with all field-level extension agents across Malawi to conduct a demonstration of fertilizer response in hybrid maize on farmers' fields. Each extension agent worked with a local farmer to plant small plots of hybrid maize to which different amounts of fertilizer were applied. The extension agent monitored the plots, collecting data throughout the growing season and at harvest. These data were used in an analysis to develop area-specific fertilizer recommendations for hybrid maize grown under smallholder farming conditions in Malawi. The aim of this research was to assess what response levels could be obtained on farmers' fields under good management.

Table 3.3 shows the hybrid maize yield response to the application of inorganic nitrogen fertilizer, together with smaller amounts of sulfur and phosphorus, obtained across the two years of this demonstration program. The maize yield response patterns shown in Table 3.3 generally make sense. The highest responses are seen in the high-potential areas of the mid-altitude plateau in the Central region—Lilongwe and Kasungu agricultural development divisions (ADDs). The lowest responses are seen in the Shire Valley ADD in the far south of Malawi due to the area's rich, alluvial soils that are not as responsive to fertilizer as are upland soils, and to the area's cropping conditions, which are more variable from year to year than elsewhere in Malawi, with both floods and droughts frequently reducing the size of harvests.[4]

However, these response rates reflect the upper bounds of the responses that can be obtained under smallholder cropping conditions. The 1995/96 and 1997/98 nationwide trials were designed to reflect the potential maize productivity levels that smallholder farmers could achieve using fertilizer under good crop management and on reasonably good soils. The design of the trials necessarily sought to eliminate many of the crop management, plot siting, and pest and disease factors that might significantly reduce maize yield response to fertilizer. The protocols for implementing the demonstrations instructed the extension agent to work with an experienced farmer. That farmer was to identify an area of his or her fields that was as optimal as possible for maize production; for example, no evidence of termite or *Striga* (witchweed) infestation, not subject to flooding, not exposed to theft or wild animals, and so on. Moreover, the crop was to be managed reasonably closely, with timely planting, timely application of fertilizer in two doses, and at least two weedings.

4 In addition to these spatial patterns, the maize yield response to fertilizer declines at higher rates of fertilizer application, a pattern which is typically found with fertilizer application.

TABLE 3.3 Average hybrid maize yield response to inorganic fertilizer applied to plots on farmers' fields in Malawi, 1995/96 and 1997/98 cropping seasons, by Agricultural Development Division

Agricultural Development Division	Maize grain yield (adjusted), kg/ha				Nitrogen (N) response, kg maize/kg N applied			Overall fertilizer response, kg maize/kg fertilizer, average	Number of sites	
	Application rate, kg/ha	35:10:0+2S	69:21:0+4S	92:21:0+4S	35:10:0+2S	69:21:0+4S	92:21:0+4S		1995/96	1997/98
MALAWI	1,281	2,194	2,719	3,041	26.1	20.9	19.1	7.8	1,680	1,408
Karonga ADD	1,470	2,270	2,677	2,899	22.9	17.5	15.5	6.6	116	106
Mzuzu ADD	998	1,873	2,488	2,853	25.0	21.6	20.2	7.9	250	217
Kasungu ADD	1,197	2,163	2,780	3,092	27.6	22.9	20.6	8.4	211	186
Lilongwe ADD	1,273	2,292	2,847	3,188	29.1	22.8	20.8	8.6	312	255
Salima ADD	1,375	2,290	2,838	3,179	26.1	21.2	19.6	7.9	130	102
Machinga ADD	1,124	2,073	2,577	2,887	27.1	21.0	19.2	7.9	280	208
Blantyre ADD	1,427	2,358	2,907	3,252	26.6	21.4	19.8	8.0	241	206
Shire Valley ADD	1,713	2,393	2,636	2,904	19.4	13.4	13.0	5.4	140	128

Source: Adapted from Benson (1999).

Note: Fertilizer application rates are based on the application of urea (46:0:0 (N:P_{205}:K_{20})) and 23:21:0+4S fertilizers in units of 50 kg bags to a 1 ha area—that is, 35:10:0+2S, 1 bag of each for 100 kg/ha total; 69:21:0+4S, 2 bags of each for 200 kg/ha total; and 92:21:0+4S, 3 bags urea and 2 bags 23:21:0+4S, for 250 kg/ha total. Response rates are calculated as (maize grain yield at level of fertilizer applied – yield without fertilizer) per kilogram of nitrogen applied for the nitrogen response, or per kilogram of fertilizer applied for the overall fertilizer response. Measured plot yields were adjusted downward by 26 percent to better reflect likely yields on farmers' fields under similar good, but not so intensive, agronomic management and to account for the high moisture content in the recently harvested grain when it was weighed. The data labeled "response, kg maize/kg N applied" do not reflect exclusively a response in grain yield to the application of nitrogen, but also responses to phosphorus and sulfur. As such, the figures overestimate the response to nitrogen to a small degree. ADD = Agricultural Development Division (subnational agricultural extension planning unit).

The maize yield responses to fertilizer presented in Table 3.3 are considerably higher than those estimated from data collected through farm household surveys, particularly those that have been done as part of evaluations of FISP. Maize yield response to fertilizer is central to any cost-benefit analysis of FISP because the yield response is the basis for the benefits of the program against which the costs of providing the subsidized fertilizer and improved seed are evaluated. Drawing upon three years of data collected from just under 3,000 households over three cropping seasons in which FISP or the earlier Targeted Input Subsidy Program was implemented, Lunduka, Ricker-Gilbert, and Fisher (2013) computed a response rate of 2.71 kg of grain per kilogram of subsidized fertilizer applied—a response rate just over one-third of that observed in the 1995/96 and 1997/98 nationwide trials (Table 3.3). Although the authors found considerable variation across farm households in response rates, using this mean response level with commercial maize prices to gauge the benefits of FISP against program costs over its first four years, they found that the costs of the program were greater than the benefits obtained in three of the four years.

In a review of input subsidy programs across 10 African countries including Malawi, Jayne and colleagues (2018) found a significant difference between the rates of response to fertilizer that farmers *may* get, based on maize fertilizer trials on-farm, and those that they *do* get, based on farm household surveys. The authors found that the on-farm fertilizer trials in the 10 countries showed yield responses equivalent to between 6 and 13 kg of maize per kilogram of fertilizer, whereas farm household surveys demonstrated much lower efficiencies, of between 2 and 9 kg of maize per kilogram of fertilizer applied, with most farm households obtaining less than 5 kg of maize per kilogram of fertilizer. This substantial difference between actual and attainable maize yield responses to fertilizer is due to the various resource constraints that smallholder farming households face, including access to quality seed and fertilizer in sufficient amounts before the time of planting, control of pests and disease, timely and adequate crop management operations, and information on making optimal use of fertilizer on their crops under the particular growing conditions they face. Malawi is not unique in this regard.

Nonetheless, the response levels computed from the 1995/96 and 1997/98 nationwide trials in Malawi indicate the levels of maize yield response to inorganic fertilizer that are potentially achievable by those commercially oriented smallholder farming households that are able to surmount the various resource constraints enumerated.[5] Farmers who are experienced and have good farming knowledge; have access to reasonably good land; obtain hybrid seed and inorganic fertilizer before the planting rains come; can manage the weeds, pests, and diseases that threaten their maize crop; and are able to apply labor to the crop when required as it grows should be able to obtain yield responses from their use of inorganic fertilizer that approach those shown in Table 3.3. Although many farm households in Malawi cannot now combine all of these factors to generate these maize yield responses from their fertilizer, responses that approach such levels should be expected by commercially oriented farmers seeking to maximize their returns from investment in inorganic fertilizer.

As a target for initiatives to improve maize productivity in Malawi, a doubling or more of maize yield responses obtained by smallholder farmers should be possible. This would require significant investments in both public and private agricultural extension services and strengthened input supply systems. And even then, it cannot be assumed that it would make financial sense for commercially oriented smallholder farmers to apply fertilizer to their maize if the costs of fertilizer are high relative to the price they will receive for the maize they produce using that fertilizer. Much more reliable maize and agricultural input markets in Malawi are needed. The weak performance of Malawi's maize markets is discussed in the next chapter.

Organic approaches to sustainable soil fertility management and intensification of smallholder cropping systems in Malawi

Given the substantial financial costs that smallholder farming households must shoulder to use commercial inorganic fertilizer, considerable research has been done in Malawi since the late 1980s exploring how smallholder farming households might substantially increase their use of organic sources of plant nutrients,

---

5 However, in using these results from 25 years ago to guide the design of policy or programs today or to assess the likely impact of inorganic fertilizer use on national or household food security, changes in aspects of fertilized hybrid maize production under good management in Malawi should be recognized. These changes might include the following:

- Improvements in the performance of the hybrid maize varieties smallholder farming households plant. These might include varieties with improved nutrient use efficiency or those that are better able to manage water stress, seasonal changes in the length of the growing season, or increased (or new) pest and disease pressures.
- Increased variability in rainfall due to climate change.
- Increased significance of deficiencies of plant nutrients other than nitrogen, phosphorus, or sulfur, such as zinc or potassium, that limit maize productivity. Deficiencies in these nutrients may emerge through continuous cropping without nutrient inputs, which depletes soil nutrient stocks, or through soil erosion.

particularly from nitrogen-fixing legumes, to increase crop yields, particularly for maize. Among the approaches that have been examined are:

- Rotations with grain legume crops, particularly groundnut and soyabean (Gilbert 2004; Bezner-Kerr et al. 2007; Chimonyo, Snapp, and Chikowo 2019)
- Intercropping maize and shorter-season grain legumes with longer-duration leguminous species, particularly pigeonpea (Adu-Gyamfi et al. 2007; Smith et al. 2016; Chikowo et al. 2020)
- Green manure crops, particularly *Mucuna pruriens* (velvet bean), with incorporation of the biomass produced (Gilbert 2004; Robertson et al. 2005)
- The traditional system of planting annual crops under *Faidherbia albida* trees (Saka et al. 1994)
- Relay cropping or row intercropping (alley cropping) of maize with various leguminous shrubs, including *Sesbania*, *Gliricidia*, *Leucaena*, and *Senna* species, among others, again with incorporation into the soil of the biomass from the shrubs (Jones et al. 1996; Phiri, Snapp, and Kanyama-Phiri 1999; Ikerra et al. 2001; Akinnifesi et al. 2008; Coulibaly et al. 2017)

A central element of all this research, done both on-farm and on-station, has involved comparing the performance of the organic technologies against that of control plots of maize without fertilizer and maize with fertilizer at recommended application rates. These comparisons have focused particularly on the effect of these technologies on the maize yield, paying less attention to several other benefits of organic soil fertility management approaches. Among these benefits are, most notably, improved soil health (primarily due to increased soil organic matter of higher quality) compared with what can be realized with inorganic fertilizer alone, and when grain legumes are a component of the technology, crop and dietary diversity. Most of the technologies, particularly when employed with close attention to management of the biomass produced by the nitrogen-fixing species involved, provide significantly higher maize yields than those obtained at the same site from the unfertilized maize control plots. The nitrogen and organic matter that they add to the farm plots generally result in increased yields of maize and other crops in such cropping systems, with these yield benefits increasing somewhat over time. If employed in an attentive manner and if their use by farming households is not constrained by labor or land shortages, the technologies can contribute to increased maize production.

However, the organic approaches do not consistently provide higher yields than are observed on the fertilized maize control plots, with most research showing that fertilized maize yields are statistically significantly higher than yields of maize grown using the organic systems without fertilizer. This is not surprising. First, the amount of nitrogen that is fixed annually by the legumes generally is less than the recommend nitrogen application rates for maize. Second, biomass management is critical to the performance of these organic technologies—if the biomass is not quickly incorporated into the soil, its quality for soil amendment purposes degrades. Unincorporated residues are often grazed by livestock or burned in preparing for the next crop, and thus supply virtually no nutrients or organic matter to the soil. In addition, with green manure crops or row intercropping within the dominant hoe cultivation systems smallholder farmers employ, incorporating the residue requires considerable labor. As with the use of inorganic fertilizer, making profitable use of these organic technologies to increase the productivity of their crops requires that farmers have considerable knowledge and experience.

Over the past 15 years, conservation agriculture farming systems have been widely promoted and researched in Malawi. These involve a combination of several different land and crop management approaches, including limited tillage, the maintenance of soil cover, and diverse crop mixes with intercropping or crop rotations that include legumes. Such systems can improve soil quality through nitrogen fixation and increased incorporation of organic matter, suppress weed growth, protect the soil from erosion, and improve water and nutrient use. Although inorganic fertilizer certainly can be a part of such systems, organic soil fertility management approaches are more commonly used. These include some of the organic approaches to sustainable soil fertility management noted. Compared with unfertilized maize, conservation agriculture systems generally have been found to improve agronomic performance after several years of implementation. However, the economic performance of conservation agriculture systems as a whole is more problematic (Jew et al. 2020), and adoption rates are low even following promotion efforts (Thierfelder et al. 2017; Chinseu, Stringer, and Dougill 2019). The high labor requirements for implementing some components, the multiyear time horizon to realize productivity benefits,[6] and the need for location-specific modifications to the set of practices used have all been identified as significant barriers to farmer uptake of conservation agriculture (Giller et al. 2009; Corbeels et al. 2014).

Consideration of how smallholder farmers in Malawi might best deploy both organic soil fertility management technologies and inorganic fertilizer is pertinent to global discussions around sustainable intensification of agricultural systems that have been going on over the past 10 to 15 years. "Sustainable intensification" is defined as the adoption of processes that increase agricultural yields without adverse environmental impact and without bringing more land into production (Royal Society 2009). While this concept says nothing about how it will be achieved (Pretty and Bharucha 2014), in practice sustainable intensification has most commonly been advocated in discussions around safeguarding the provision of the full set of environmental services wherever agriculture is practiced, going beyond sustainable crop yields alone. Consequently, organic approaches generally are privileged in discussing how soil fertility management practices might contribute to sustainable intensification processes.[7]

However, given the significant impact of smallholder productivity levels on national food security, raising crop yields is the most important goal for agricultural transformation in Malawi. Better yields are also critical to achieving other sustainable intensification goals within Malawi's agriculture sector. However, organic approaches to soil fertility management alone are unlikely to sufficiently raise crop yields in smallholder farming systems in Malawi. Making efficient and profitable use of these organic approaches for field crops requires farmers to surmount significant knowledge barriers and make often limited labor available at specific times. Without site-specific knowledge built through farmer experience and experimentation or obtained through advice from agricultural extension experts, the risk is high that adopting farmers

6 Particularly green manure crops and row intercropping require that farmers accept the proposition that in the medium term they will receive more maize or other food crop production in aggregate by devoting all of a plot for a season (green manure) or part of their plot continuously (row intercropping) to complementary crops than they would if they used the plot solely for maize or other food crops. Given the high risk of food insecurity that many farm households in Malawi face, their limited landholdings, and the weak markets upon which they would need to rely for food in the interim, most choose not to dedicate some of their cropland to these types of organic systems of soil fertility management that interrupt the supply of maize from their farm plots.

7 There is debate in the agricultural and related sciences literature about the definition and scope of sustainable intensification, with many commentators finding the concept highly ambiguous and one that can be used to justify any agricultural development goal. See Tittonell (2014) and Petersen and Snapp (2015), among others.

will realize poor crop harvests (Jayne et al. 2019). With insufficient labor, farm households may find that the labor they did dedicate to these organic approaches could have been more profitably employed on other tasks, whether within or outside of agriculture.

The situation with inorganic fertilizer is similar. As discussed, most smallholder farming households across Malawi realize far lower maize yield responses than they could achieve if they could make more effective use of the input. Among the barriers they must surmount are difficulties in accessing information on and building experience with how best to use the fertilizer they obtain for the specific crops they grow on the particular soils they farm (Kopper, Jayne, and Snapp 2020), with the added challenge—insurmountable for many farm households given their income levels—of accessing commercial fertilizer.

The two approaches to soil fertility management—organic or inorganic—are not mutually exclusive. Although both are problematic for many farming households to employ, both are needed for any sustainable intensification of smallholder agricultural production in Malawi.

Fertilizer subsidies

The potentially achievable maize yield responses to the application of inorganic fertilizers justify the government's repeated focus on increasing access to inorganic fertilizer as both a food security strategy and an agricultural development strategy for Malawi. Even if the average farmer cannot achieve the response rates obtained in the national demonstrations (more than 7 kg of maize per kilogram of fertilizer applied), achieving even half that response is attractive from a food security perspective. If managing any food crisis resulting from low production might require imports of maize grain, most policymakers accept the argument that it makes more sense to import one bag of fertilizer rather than import the three or four bags of maize that were not produced because a farmer did not have access to that fertilizer.

However, for many farmers, the high cost of inorganic fertilizer prevents them from increasing their maize yields. Although credit could allow farm households to purchase inorganic fertilizer when they do not have income from other sources, credit is not readily available for maize production in Malawi. Given the uncertainty of rainfed agriculture, the thin maize markets, the consequent significant volatility in maize prices, and the restricted asset base of most smallholder farmers, formal credit is effectively not available. Currently no formal credit institutions primarily oriented toward agriculture operate in Malawi at a large scale.

Consequently, supplying fertilizer directly to farmers at significantly below cost has been the solution repeatedly adopted by the government to assure sufficient maize production in the country. Fertilizer subsidies were part of the policy of the colonial government in the post–World War II period and that of the first president of independent Malawi, Kamuzu Banda. The Muluzi government, the first in Malawi's multiparty era, ended fertilizer subsidies in the mid-1990s as part of a market liberalization package advocated by Malawi's development partners. However, with declining maize production and an emerging food crisis, the Starter Pack input subsidy program was implemented in the late 1990s, which provided all smallholder farmers for free with small packs of fertilizer and improved seed sufficient to plant 0.1 ha. Then, following a food crisis in 2005, the Bingu wa Mutharika government introduced a significantly scaled-up program, FISP. This initially targeted about half of all smallholder farming households in Malawi to provide them

BOX 3.1 Justification for the scale of Malawi's Farm Input Subsidy Program

In Malawi, it is estimated that at least 80 kg of plant nutrients are lost from each hectare planted in crops each year through harvest, removal of remaining crop residues, and erosion (Elwell and Rook 1996; Henao and Baanante 1999). Organic sources, including manure and leguminous crop residues, are estimated on average to annually replenish 8 kg, resulting in a net loss of 72 kg of crop nutrients per hectare. The Farm Input Subsidy Program (FISP) package for maize includes 50 kg of the principal basal fertilizer (23:21:0+4S—applied immediately after plant emergence) and 50 kg of the main top-dressing fertilizer (urea 46:0:0—applied four weeks after basal fertilizer application). This provides 46 kg of plant nutrients on the average agricultural landholding of 0.65 ha and allows production levels of around 1.75 tons/ha to be maintained, assuming that each kilogram of fertilizer will result in 5 kg of additional grain above a base unfertilized maize yield of 1.0 tons/ha.

However, the higher national maize production levels needed to meet the staple food requirements of Malawi's growing population and to contribute to economic structural transformation will result in increased annual crop nutrient losses through harvesting and, consequently, require higher levels of fertilizer application to sustain. The National Agriculture Policy (Malawi, MoAIWD 2016b) sets a maize production target for the country of 4.0 tons/ha. The 2.25 tons/ha increase in average maize yields sought above those that can be attained under the FISP package described will require the additional application of between 120 and 200 kg of nitrogen per hectare. To sustainably achieve 4.0 tons/ha maize yields, the application per hectare of a package of 3 bags of 23:21:0+4S, 50 kg each, and 6–10 bags of urea, depending on the with sufficient inputs to produce 0.5 ha of fertilized maize, with beneficiaries responsible for paying only about 35 percent of the cost. The technical arguments for the design of FISP and the size of the inputs package offered to beneficiaries are sketched out in Box 3.1. Although there were reforms to the FISP between 2014 and 2020 under the A. Peter Mutharika government, which involved some reduction in its scope and in the size of the subsidy provided to beneficiaries, following the inauguration of the new Chakwera government, an expanded program, the Affordable Inputs Programme, was instituted for the 2020/21 cropping system.

The core of FISP is the provision of 100 kg of fertilizer and 10 kg of hybrid maize seed to beneficiaries, inputs generally sufficient to produce 0.5 ha of maize. The principal beneficiary selection criterion established early in the program was that recipients be resource-poor (unable to use their own resources to access unsubsidized inputs) but productive farmers. Excluded

fertilizer use efficiency the farmer attains, would be required. These are fertilizer application levels comparable to those used in South and East Asia, where there has been significant growth in crop yields over the past 50 years resulting in part from use of heavily subsidized fertilizer.

In the context of Africa, the FISP fertilizer package is considered ambitious. However, although there has been considerable discussion of the fiscal sustainability of FISP and recurrent calls for a much smaller program, the agronomic evidence on the nutrient inputs required to sustain production levels that meet maize demand in Malawi shows that a larger-scale fertilizer subsidy program could be justified in the absence of other efforts to make full-market-cost fertilizer use on maize a profitable financial proposition. Farmers need to be in a position to access and efficiently use inorganic fertilizer to manage soil fertility on their fields in a manner that replenishes annual nutrient losses and propels maize productivity toward the government's target of average maize yields of 4.0 tons/ha. Particularly if the fertilizer is used significantly more efficiently by beneficiaries, an enlarged fertilizer subsidy program would enable them to reach this target.

However, an enlarged program would exacerbate the important opportunity costs that Malawi now is bearing with FISP in terms of addressing the many other development challenges the country faces. Moreover, further entrenchment of FISP will forestall needed market reforms and other efforts necessary to make the use of fertilizer at full market cost a profitable financial proposition for smallholder maize farmers.

Source: Adapted from Carr (2014).

from the group of FISP beneficiaries were both the poorest and wealthiest 20 percent of farm households, with the program targeted to reach the 60 percent of smallholder farming households in the middle. Community targeting involving local traditional authorities and community committees has been used to identify beneficiaries. Later, considerations of vulnerable households were brought into the targeting decision, adding social protection objectives to the initial production focus of FISP. Beneficiaries are provided vouchers that they can use to obtain fertilizer and seed from private agrodealers at a subsidized price.

The most complete information on the scale of FISP is for the period covering the 2005/06 to 2011/12 cropping seasons (Chirwa and Dorward 2013). Over this period, the program reached between 50 and 80 percent of smallholder farming households, depending on the year—averaging 1.6 million beneficiary households. Most assessments of the performance of the program find that it has been poorly targeted, with smallholders across all welfare levels accessing the input subsidies (Nkhoma 2018). On average, 165,000 tons of subsidized fertilizer were sold each season. The overall subsidy received by farmers ranged from 64 to 93 percent of the market price for the inputs. The average net cost borne by government per year for FISP over this period was just over $115 million, but this rose to $214 million in the 2008/09 program year due to a spike in fertilizer prices on international markets. Annual costs between the 2005/06 and 2011/12 cropping seasons represented an average of 55 percent of the government's annual budget for agriculture, 8.8 percent of the total government budget, and 3.5 percent of Malawi's GDP. The input subsidy program remains the largest government project in Malawi's agriculture sector.

FISP has contributed to improving national food security simply through increasing maize production. However, all independent assessments conclude that the program could be implemented in a more efficient manner and achieve significantly broader impact, particularly at farm level. As discussed above, the levels of maize yield response to fertilizer found in evaluations of FISP are significantly lower than what can be achieved on smallholder maize plots in Malawi under good management. The opportunity costs associated with this inefficiency in program implementation at farm level are significant. Lunduka, Ricker-Gilbert, and Fisher (2013) computed the benefit-to-cost ratios for the program to be less than 1.0, meaning that costs exceeded benefits, in three of the four years of program implementation examined (2005/06 to 2008/09). Using more optimistic estimates of the maize yield response to fertilizer on the plots of FISP beneficiaries, Chirwa and Dorward (2013) estimated an overall benefit-to-cost ratio of 1.09 for FISP over the six years from 2006/07 to 2010/11—somewhat greater direct benefits than costs.

When indirect, second-round economic effects of FISP are brought into the computation of benefits, larger positive net benefits are obtained. These benefits are linked to the increased economic activity, lower food prices, and increased demand for labor that the program fosters. Using a computable general equilibrium model of Malawi's economy, Arndt, Pauw, and Thurlow (2016) calculated that these indirect benefits increase the total benefits that can be attributed to FISP by about 60 percent, boosting their estimate of the benefit-to-cost ratio of the program from 0.99 (meaning that costs slightly exceeded direct benefits) to 1.62 when indirect benefits are included.

However, there is no strong evidence that even when these second-round benefits of FISP are included, the impact of the program on the well-being of farm households, the development of the agriculture sector, or national food security is superior to that which might be achieved through allocating the resources put into FISP toward other immediate objectives, such as improved rural transport and communication infrastructure, agricultural research, or improved advisory services for farmers, among others. The high costs, obvious technical inefficiencies, and some evidence of politicization in the implementation of the FISP program have led to increased questioning of whether this is the best use of the significant public resources allocated to the program. Recognizing the potentially important opportunity costs for the overall development of Malawi associated with the inefficient implementation of FISP, coupled with a fiscal crisis for the government linked to withdrawal of donor financial support due to concerns about mismanagement, the government made reforms to the program starting in 2015/16 (Nkhoma, Bosman, and Eduful 2019). These included reducing the number of annual beneficiaries, from around 1.5 million farming households to 900,000, as well as the amount of subsidy farmers receive. These changes reduced the cost of the program significantly—the average annual government budget for FISP for the 2016/17 to 2018/19 seasons was just under $50 million, or less than half of the average annual net costs incurred by the program between 2005/06 and 2011/12.

But Malawi's political leaders have little interest in eliminating such input subsidy programs altogether. Absent compelling and broadly recognized reasons for doing so, sharply downscaling or eliminating the fertilizer subsidy program would weaken their political support (Dionne and Horowitz 2016; Nkhoma 2018). In the presidential campaign in early 2019, all major political parties confirmed that they would, if their candidate were elected, maintain fertilizer subsidies for farmers, and several parties stated that they would make the subsidy program universal with only minimal targeting. For the 2020/21 cropping season, the new Chakwera administration implemented an expansion of the FISP, the Affordable Inputs Programme, that provided subsidized inorganic fertilizer and cereal seed to 3.5 million smallholder farming households.

In the short term, the principal justification for such input subsidy programs is food security. Given its political importance, any plans to eliminate or sharply reduce the current scale of the input subsidy program would face significant political challenges. Increasing political support for a significant reduction in the scale of the input subsidy program would require a clear demonstration of how the smaller modified program would improve access to food for Malawi's vulnerable households.

The cost of FISP per delivered ton of fertilizer in 2011/12 was about $950. The costs to the World Food Programme of delivering 1 ton of imported maize to beneficiaries in rural Malawi are reportedly about $600, based on maize import parity prices in the range of $350 to $450 (WFP 2014). Assuming 5 kg of maize output per kilogram of fertilizer applied, it would cost $3,000 to import the amount of maize equivalent to what 1 ton of fertilizer supplied through FISP would produce. The economics of fertilized maize production are driven by the price ratio of fertilizer to maize, and the agronomic efficiency through which farmers convert the nutrients in the fertilizer into grain. Under current price patterns and farming systems, fertilizer subsidies remain the principal way in the near term for the government of Malawi to assure national food security, reduce its dependence on international humanitarian assistance for famine relief, and (partially) address degradation of the soils on which the country depends for its food. However, a fertilizer subsidy program is not a strategy for sustainable agriculture-sector growth or for comprehensively addressing chronic food insecurity.

Sharply scaling back the government's input subsidies to engage in a program of more sustainable agricultural development that will better ensure Malawi's food security will be challenging. Such a program would entail higher, but overall less volatile, maize prices to provide incentives for increased commercial production that would be channeled to consumers through more reliable maize markets. Given the price shocks this would entail to the welfare of many households, increased food aid and almost certainly an expansion in social protection programs would be required during the transition, even as significantly larger investments are being made in agricultural research, in expanded extension services, and in market and transportation infrastructure.

However, it also is important to recognize that fertilizer subsidies as a strategy to improve food security are dependent upon good rains. Food aid is generally not needed in Malawi following good rainy seasons. Since the 2012/13 cropping season, each year Malawi has seen large numbers of people vulnerable to food insecurity due to local droughts or floods, particularly in the Southern Region of the country, and significant efforts have been required to meet their food needs. This has been the case despite the $50 million or more spent annually by the government to support FISP. Although input subsidy programs may help in bridging structural food deficits in years of normal rainfall, such programs are not an effective way to manage the acute food insecurity that arises under poorer cropping conditions.

Increasing commercial purchases of inorganic fertilizer by smallholders should be the longer-term objective if use of fertilizer is to contribute to increased maize production for food security. However, doing so is problematic. All economic analyses over the past 20 years have demonstrated that there is little financial sense in a farmer using fertilizer on maize produced solely for commercial sale, and not for home consumption (Benson 1999; Darko et al. 2016). Malawian farmers cannot purchase the fertilizer they need with confidence that they will be able to pay for its full cost out of the proceeds from the sale of the maize produced. The price of maize, in real terms, will need to make commercial fertilized maize production profitable for Malawian farmers. But of course, any efforts by government to increase real prices for maize in order to improve the economics of fertilized maize production would almost certainly face intense political opposition given that the majority of Malawians are net purchasers of maize. However, viable alternatives to subsidizing fertilizer to maintain higher levels of maize production in Malawi in the short term are not obvious. As will be discussed in the next chapter, critical reforms are needed to improve the performance of Malawi's maize markets and to increase price stability in those markets before commercial production of fertilized maize without any fertilizer subsidies will be a significant component of the food security of the country.

Moreover, whereas the large fertilizer subsidy programs that have been implemented in Malawi off and on since the mid-1990s often have been justified in other terms, including to allow farmers to experiment with commercial inputs and to provide a social safety net for vulnerable rural households, the political rationale for these subsidies has been to ensure national food security and the household food security of beneficiaries. In their comprehensive review of the experience of Malawi with fertilizer subsidies, Chirwa and Dorward conclude that a principal, if not the primary, driver for their use is "widespread understandings among the Malawian population that fertilizers are critical to food security, that this is dependent on food self-sufficiency, and that the government has an active responsibility in ensuring food self-sufficiency and hence in enabling widespread fertilizer access" (2013, 70).

Public resources are used to overcome the problematic economics of fertilized maize production in Malawi. If conditions are right, the short-term food security of the nation may be bolstered in doing so. However, this has been done at the expense of the medium-term development of deeper agricultural markets for inputs, outputs, and credit. Consequently, it is difficult to see an easy exit from a continued reliance on large fertilizer subsidies if Malawi's markets remain underdeveloped. Strong markets are needed to effectively induce sufficient production of maize and other crops by Malawi's farmers to meet the consumption needs of Malawi's consumers at reasonable prices.

Irrigation

Despite significant surface and groundwater resources, most of Malawi's agricultural production is rainfed—only about 10 percent of the maize produced in the country between 2013/14 and 2017/18 came from irrigated plots (Table 3.2). Almost all the rain that Malawi receives falls over the five months from December to April, with most fields left uncropped throughout the rest of the year. Although substantial investments have been made in engineered irrigation schemes both for large-scale commercial purposes, particularly sugar production, and for smallholders, most irrigated crop production by smallholders generally involves very small-scale production in the dry season in local *dambo* wetlands and along the banks of streams using manual water supply systems to supplement residual soil moisture.

Of the 2.5 million ha of land estimated as being farmed in Malawi, just over 100,000 ha was estimated to be irrigated in 2015. Of this area, 37,000 ha is farmed by smallholders using manual water supply methods; just under 20,000 ha is farmed by smallholders in engineered irrigation schemes, most of which are small; and the balance of 48,000 ha of irrigated land is found on estate lands, with the two principal sugar estates in Malawi accounting for more than 70 percent of estate land under irrigation (SMEC 2015a). In 2011, it was estimated that only 0.5 percent of all crop plots in the country were irrigated (Malawi, NSO 2012). The analysis for the 2015 National Irrigation Master Plan and Investment Framework assessed the potential irrigation area for the country, based on available water, at just over 400,000 ha. However, irrigation development is capital intensive. The average cost of development for the 43 new projects prioritized under the master plan is projected to be $5,300 per hectare (SMEC 2015a).[8]

Most analysts recognize the potential for expansion of irrigated farming in Malawi to contribute to increased food crop production, significant increases in income (SMEC 2015b), and enhanced welfare for the households involved (Benson 2015). However, most assessments also highlight that formal smallholder irrigation schemes, in particular—as opposed to informal irrigated plots operated on an individual basis or irrigation on commercial estates—cannot be operated profitably and sustainably without close management and clear understanding of the economic opportunities that irrigated farming in Malawi offers, as well as the limits on those opportunities. The economics of sustainable irrigated farming in Malawi are challenging. To be profitable, whatever crops are grown must provide sufficient returns to cover the operating and maintenance costs of the irrigation infrastructure, the costs associated with effective scheme management in communal irrigation schemes, and the basic costs of production and marketing. Profitability will be determined by the choice of crops, the productivity of irrigated farming, the number of crops that can be harvested annually, and the access that farmers have to both the commercial inputs they require and sufficiently large output markets (SMEC 2015b).

In general, staple food crops have not been a good choice for smallholder irrigation schemes in Africa (Inocencio et al. 2007). In the case of Malawi, the price of maize grain is determined by the supply of rainfed maize, which under good management and with good rains can attain productivity levels similar to those of irrigated maize. Production of irrigated maize for grain is financially advantageous only in poor-rainfall seasons, which remain difficult to predict, or when grown in the dry season following a poor rainfed season for maize production nationally. More reliably profitable crops are higher-value crops whose supply is limited in the dry season. For smallholder farmers, these crops are primarily vegetables, including green maize, for sale in urban centers and towns. However, because such crops are perishable, the location of the irrigation scheme then becomes a significant consideration, with farmers in remote schemes facing significant costs in delivering such produce to market in salable condition. What is going to be produced, whether there is sufficient market to absorb what is produced, and how the produce will be marketed

8 This estimated per-hectare cost for the development of new irrigation systems is much lower than other estimates. Inocencio et al. (2007) estimated an average cost per hectare of $14,450 for the 26 systems in Africa south of the Sahara involving new construction that they reviewed.

are important considerations in the design and sustainability of any irrigation investments in Malawi (SMEC 2015b).

The idea that Malawi's vast water resources can be used to irrigate the food crops the country requires is appealing and has driven substantial investment in irrigation development, both by Malawi's development partners and by the government. Since 2010, the government has been implementing the Green Belt Initiative to expand irrigated farming in the country, ostensibly to reduce the vulnerability of rainfed food production, and hence of food availability, to poor weather conditions.[9] However, basic economic considerations show that smallholder sustainable irrigation to produce staple crops, including maize, will not be economically viable, and so will not make a direct contribution to improving the country's food security on a large scale. Although irrigating farmers are likely to see important food security benefits from their use of irrigation, these benefits will be realized primarily indirectly, as the farmers use income from irrigated production of commercial nonstaple crops to purchase staple foods. Markets for high-value food crops are the key channel for translating smallholders' access to irrigation into improved food security in Malawi. It is not about smallholders directly producing irrigated maize to supply the staple grain stocks that the nation needs.

In addition to choosing the right crops, both internal and external organizational challenges need to be addressed for production from smallholder irrigation schemes to be sustainable. Internally, effective scheme management is required so that water is distributed fairly and the costs for water used are recovered from farmers in the scheme. Such management capabilities are not readily found in rural communities in Malawi because they require both trained leadership and community consensus. Training and institutional capacity-building services need to accompany any new irrigation scheme.

Similarly, the sustainability of such irrigation schemes depends on prudent management of local resources within the watershed feeding each scheme, including management of local soil resources to reduce erosion and siltation. Consequently, watershed or groundwater management mechanisms should be set up to govern how all users of those resources, both irrigating farmers in the scheme and residents in neighboring communities, employ them in a sustainable and fair manner. Here too, training and institutional capacity-building

9 However, despite justifying the establishment of the Green Belt Initiative on food security grounds, the principal efforts of the agency have been focused on expanding irrigated sugarcane production by developing a new sugar plantation in Salima district (Chinsinga 2016). This site began production in 2017 and is primarily a vertically integrated facility, growing most of the cane it processes, although it does obtain some cane from smallholder outgrowers.

services are required, as well as acceptable mechanisms for resolving any conflicts that may arise over water use.

Irrigated farming can contribute in important ways to the growth of Malawi's economy, to supplying consumer markets with a more diverse range of food crops, and to improving the welfare of those engaged in such farming. However, it is not an easy solution to adopt, requiring significant technical analysis and expertise, capital, and effective market linkages. Nonetheless, that only about one-quarter of the potential irrigation area for the country is now exploited means there is a large investment opportunity through which agriculture can make important contributions to Malawi's economic growth.

Access to Agricultural Land

Land is critical to agricultural production. Here some aspects of access to agricultural land in Malawi are considered. First, the customary land tenure system under which most farming households obtain land to use is described, with specific attention to landlessness within rural communities. The discussion then examines the distinction between the smallholder and estate subsectors that together compose the agriculture sector in Malawi. The smallholder sector is typified by small landholdings obtained under customary tenure on which crops are grown primarily for consumption by the farming household. In contrast, estates generally involve larger areas of land obtained through freehold ownership or through long-term leasehold arrangements, on which crops are produced for sale, particularly into Malawi's export markets. This section assesses whether this qualitative distinction between smallholders and estates is now of much value for guiding agricultural development efforts.

Customary access

Most farmers in Malawi have access to the land that they farm under customary tenure. They have rights to the use of this land by virtue of their being members of a local lineage, with the jurisdiction over such land rights typically vested in local traditional authorities. To an important degree, these customary rights to land provide a safety net for most Malawian households, permitting them to engage in subsistence production if other economic options that they might pursue are not sufficiently remunerative (Ellis 2005). However, because most households farming under smallholder conditions are capital and credit constrained, and thus unable to use modern inputs and technologies on their crops, productivity levels are generally low for the farmland under customary tenure.

Customary rights to land are primarily usufruct rights. Nonetheless, with the grantee or his or her family members or descendants making regular use of the land, once the land is allocated, it is considered to be the property of the grantee or his or her family (Chirwa 2008). However, no title is provided to the users and no cadastral record of the land allocation is made. Although this system allows most Malawian households to have access to farmland, generally only small plots of land can be obtained in this manner. Based on

TABLE 3.4 Characteristics of rural households without and with agricultural land, 2016/17

Characteristic						
All households	National		Rural		Urban	
Households without agricultural land in 2016/17, %	21.0	(0.81)	10.1	(0.70)	67.6	(2.78)
Households without agricultural land in 2010/11, %	16.6	(0.80)	7.7	(0.66)	64.6	(3.75)
Observations, 2016/17	*12,447*		*10,175*		*2,272*	
Rural households only	Overall		Landless		With land	
Poor households (per capita consumption below national poverty line), %	51.9	(0.87)	35.3	(2.01)	53.7	(0.87)
Wealthiest quintile of households, nationally, %	12.1	(0.52)	25.7	(1.61)	10.5	(0.48)
Household size	4.3	(0.03)	3.5	(0.08)	4.4	(0.03)
Age of household head, years	43.7	(0.21)	37.7	(0.65)	44.3	(0.23)
Female household head, %	30.8	(0.59)	24.7	(1.55)	31.5	(0.63)
Married household head, %	70.1	(0.60)	64.4	(1.75)	70.7	(0.64)
Separated or divorced household head, %	14.2	(0.42)	13.8	(1.23)	14.3	(0.45)
Widowed household head, %	13.5	(0.42)	11.5	(1.10)	13.7	(0.45)
Never-married household head, %	2.3	(0.21)	10.4	(1.18)	1.3	(0.14)
Household head born in other district than that of current residence, %	16.2	(0.68)	30.1	(2.35)	14.7	(0.62)
Household head received no education, %	20.0	(0.61)	12.9	(1.41)	20.8	(0.62)
Household head received some primary education, %	62.2	(0.64)	53.5	(1.84)	63.2	(0.65)
Household head received some secondary or tertiary education, %	17.8	(0.60)	33.6	(1.88)	16.0	(0.55)
Observations, 2016/17	*10,175*		*1,253*		*8,922*	

Source: Author's analysis of 2016/17 Malawi Integrated Household Survey 4 data (Malawi, NSO 2017), weighted. Figures in parentheses are the sample design–corrected standard errors for the estimates.

Note: "Households without agricultural land" for the survey analysis here are those that reported not owning or not otherwise cultivating a plot during both the last completed rainy and dry *(dimba)* seasons.

analysis of the 2016/17 IHS4, the mean land area cultivated by households in Malawi that grew any crops is estimated at 0.65 ha per household and 0.17 ha per household member. Between 65 and 75 percent of all land in Malawi is estimated to be under customary tenure, with the rest being either public land (15 to 20 percent) used for public purposes, including national parks and forest reserves, or land held under private tenure (10 to 15 percent) through a freehold title or through registration of previously customary land as private (USAID 2010).

Based on IHS4 estimates, the share of rural households in Malawi that do not make use of agricultural land remains quite small, at 10 percent. However, the share of households without such land is rising, up from an estimated 7.7 percent based on the previous IHS3 survey of 2010/11. A broad analysis (Table 3.4) shows that landlessness in rural Malawi is not associated with poverty or economic vulnerability.

Rather, rural households that do not have land to farm are less likely to be poor, more likely to be in the wealthiest quintile nationally, smaller in size, and less likely to be headed by females than the norm. In terms of household formation and marital status, married heads of household are unlikely to be landless. Never-married heads of household are most likely to be landless. This likely reflects both the current occupational choices of young adults—particularly those defined for the survey to be heads of household but who are still pursuing their education—and the fact that land rights within communities are not vested in unmarried young adults but remain with the previous generation until the younger one is married. Heads of household who are not natives of the district in which they live are also much more likely not to have access to agricultural land, reflecting the importance of ties to local lineages. Finally, rural households headed by relatively well-educated individuals, who are often employed in salaried positions outside of agriculture, are more likely to be landless, while it is somewhat rarer for those with only primary education or no education at all to not have access to land.

These findings suggest that workers in landless rural households pursue livelihoods that require secondary or higher levels of education and some level of mobility, given that the heads of many landless households do not reside in their district of origin. In terms of their contribution to household welfare, these livelihoods are evidently superior to any agricultural livelihoods workers in those households might pursue. Rural households headed by teachers or medical workers, for example, likely make up a disproportionate share of landless rural households.

Closer analysis of the IHS4 data shows that landless rural households are more prevalent in the Northern region, although additional research is needed to confirm this finding because it runs counter to expectations based on relative population pressure. At the district level, a somewhat higher share of landless rural households are seen in Nkhata Bay district in the Northern region; Salima and Nkhotakota in the Central region; and Blantyre, Mulanje, and Chikwawa in the Southern region. The higher levels of landlessness in some of these districts may be associated with the sugar and tea estate workers and tenants on tobacco estates.

Given the steady growth of Malawi's rural population, the continued access to land that the customary land tenure system offers members of local lineages, coupled with intergenerational land transfer patterns within families that tend to divide land among all sons or all daughters of the next generation (depending on whether patrilineal or matrilineal landholding precepts apply), it is likely that poverty is now more closely associated with near-landlessness than with actual landlessness. Several land allocation processes are probably operating in parallel, with the result that the number of rural households that do not have access to sufficient farmland to meet their members' needs is increasing.

In the context of rising population pressure and increased mobility with people changing residence after marriage, it seems certain that local land conflicts will increase over time. The informal customary system for land access can be expected to erode as competition for land increases. The poor are more likely to lose out in such conflicts, so that rural landlessness will become more strongly associated with rural poverty,[10] and the current profile of landless rural households presented in Table 3.4 can be expected to change significantly in coming years.

Decline in significance of distinction between smallholder and estate agriculture

Larger parcels of agricultural land are held under freehold or, more commonly, leasehold terms. Land under freehold is primarily a legacy of the estates that were granted in the early colonial era to colonial settlers, particularly in the Southern region, as well as smaller parcels that were provided to Malawian farmers under some land redistribution and development programs in both

10 Issues of rural landlessness and the welfare impact of insecurity in access to agricultural land in Malawi are more closely examined in Berge et al. (2014) and Deininger, Xia, and Holden (2019), together with discussions on the challenges of creating equitable systems for granting access to land in Malawi under rising population pressures and conflicts over land use.

the colonial and postcolonial periods (Ng'ong'ola 1986). Leasehold became an increasingly common means for commercial farmers to acquire land in the 1970s and 1980s when the government promoted estate agriculture, particularly for tobacco production. Public, private (freehold), and customary land all can be leased temporarily to a user, generally for 21 years, with some leases extending up to 99 years (USAID 2010).

Although corporations used leasehold arrangements to acquire large blocks of land in the 1970s, in the 1980s much smaller blocks of land were acquired by individuals, including former estate managers, businessmen, and civil servants. Between 1970 and 1995, the number of estates larger than 10 ha was estimated to have increased from fewer than 250 to around 30,000, covering more than 900,000 ha (ELUS 1997). The Estate Land Utilisation Study estimated that 67 percent of all tobacco-producing estates in 1995 were between 10 and 20 ha in size, with only 7 percent of them larger than 100 ha (Steele 1997).

There was considerable justification for distinguishing the smallholder and estate subsectors in the colonial era and during the rule of Kamuzu Banda following independence. During these periods, the state regulated the estate sector differently than the smallholder sector. Early in the colonial period, estate agriculture was seen by the state as the basis for a productive commercial agriculture sector, although this policy bias was not sustained. After independence, under Kamuzu Banda, new efforts were made to expand the estate sector (Kydd and Christiansen 1982; Pryor and Chipeta 1990). Estate producers frequently were accorded preferential access to commodity markets, whereas the Agricultural Development and Marketing Corporation (ADMARC) and earlier parastatal marketing boards controlled how smallholders sold their produce. Legally, smallholders could sell their tobacco production only to ADMARC, at significantly lower prices than those offered to estates on commodity markets. By reducing the returns to their agricultural production, the poor prices ADMARC offered smallholders resulted in a movement of labor out of the smallholder sector into wage or sharecropping employment in the estate sector, almost exclusively on tobacco estates (Kydd and Christiansen 1982).

However, with the liberalization of tobacco production in Malawi in the 1990s, the regulatory barriers to smallholders' direct participation in the national tobacco market were removed. In consequence, the profitability of estate production of air-cured burley tobacco, in particular, increasingly became problematic. Once smallholders could produce the crop on their own account, the previously dominant tenant-based burley production systems on estates were difficult to sustain. As noted, burley tobacco production has shifted almost completely from estates to smallholders since the mid-1990s.[11] Tobacco estates continue to retain a significant share of flue-cured tobacco production, however, because it is more capital intensive.

There are few economies of scale to be realized in burley tobacco production, because the crop requires close attention by the grower to ensure high-quality leaf. Mobilizing sufficient labor, not land, has proved to be the principal challenge to profitable tobacco production in Malawi (Prowse 2013). In consequence, estates find profitable production increasingly difficult when they have to offer wages that are competitive with the returns smallholders can obtain from producing their own burley tobacco. Although the production of some field crops on estates, including soyabean and seed maize, may exhibit returns to scale on the larger landholdings of estates, the markets for such crops are significantly smaller than those for tobacco. In terms of technical considerations for most crops, there is now little to distinguish smallholder production from estate production.

Similar to smallholders, estates are not engaged in a significant way in staple food crop production for market sale. The thin markets and unpredictable price patterns for maize in Malawi result in significant disincentives for both large- and small-scale farmers to engage in commercial production. Interviews in late 2015 and early 2016 with 15 large-scale agricultural producers across Malawi found that none of them engaged in commercial maize production (Edelman et al. 2016). Although all the producers surveyed reported that they grow maize for their workers or tenants to ensure that their consumption needs are met, none of them allocate land to maize for commercial sale. Because these large-scale farmers have no confidence that they will find a profitable market for maize grain or other staple food crops, they choose other, more reliably profitable crops to produce, including tobacco. The weak markets for maize in Malawi drive estates to grow maize for their staff, but to produce no more maize than that.

Concerns continue to be expressed about underused estate land. A recent project to digitize estate land boundaries for the Ministry of Lands showed that when the boundaries for the more than 24,800 estates that could be digitized were overlaid on recent satellite imagery, only 42 percent of estate land was found to be under crops (Deininger and Xia 2017). Given the challenges to engaging in high-productivity commercial agriculture in Malawi, many estate

11 This shift in burley tobacco production from estates to smallholders replicates a pattern seen in southern Malawi relatively early in the colonial period—increasing smallholder production of tobacco led to labor constraints on undercapitalized settler estates, rendering many estates financially unviable (Green 2012a).

owners are unable to profitably bring their land into production. This is not a new concern. In the 1990s, the Estate Lands Utilisation Study was specifically tasked with considering the issue of abandoned or otherwise underutilized estates. An exploratory study found that factors associated with unused estate land included small size, absentee management, lack of capital, disputes over land rights, and illness or death of the operator (Mapemba 1997).

Despite the challenges that estate holders have repeatedly faced in profitably running their larger farms, there is evidence that entrepreneurial Malawian farmers are acquiring significant areas of customary land in order to develop larger farms of between 5 and 50 ha in size. As much as 300,000 ha nationally is estimated to have been acquired by such farmers since 2005, or about 8 percent of the land under crops in the country (Anseeuw et al. 2016). Many such farmers, particularly those acquiring larger holdings of more than 30 ha, are absentee farmers with significant nonfarm income sources who reside in urban areas and, it is assumed, will invest an important share of the returns from their farming in their urban places of residence. In contrast, medium-scale farmers with smaller landholdings, of between 5 and 30 ha, are more likely to be local residents.

Although much of this land is held under informal leasehold terms, many medium-scale farmers are able to successfully remove the land to which they gain access from customary control—the parcel is ceded to the state with the consent of local authorities, and then the farmer obtains a long-term lease, generally a 21-year lease, from the government. Although bringing this land into more formal land markets through leaseholds may increase commercial agricultural production in Malawi, it is at the cost of increased risk to the livelihood security of the communities that formerly had authority over who uses the land.

However, it remains unclear whether these emerging medium-scale farmers are qualitatively different from the many farmers who acquired small estates in the 1980s. The earlier boom in the creation of small estates was an outcome of government policy that included providing prospective small estate owners with access to low-cost credit subsidized through taxation of smallholder farming (Kydd and Christiansen 1982). In contrast, the current trend in land acquisition, at least on the face of it, does not involve government action or public investment. Rather, it seems to be wholly driven by commercial decisions made either by local farmers seeking to increase the scale of their production or by urban entrepreneurs diversifying their economic activities. Nonetheless, these emerging medium-scale farmers face the same business challenges related to crop choice and capital and labor constraints on profitability that many of those who established small estates in the 1980s found difficult to overcome. As will be discussed in more detail later, these commercially oriented medium-scale farmers are likely to be an important part of any future growth in the agriculture sector. However, their contributions to agricultural transformation in Malawi will be realized only if they can sell their produce in remunerative markets, whether domestic or export.

Current Policy Framework for Agriculture in Malawi

This chapter concludes by considering the policy framework guiding public investments in the development of Malawi's agriculture sector. The overall development vision for Malawi is expressed in the Vision 2020 document of 1998[12] and the current medium-term development plan, the Malawi Growth and Development Strategy III (2017 to 2022). The National Agriculture Policy (NAP) of 2016 reflects those higher-level development policy frameworks (Malawi, MoAIWD 2016b). Encouragingly, the NAP has a stronger emphasis than earlier policies for the sector on sustainable economic growth through agriculture and treating farming as a business, rather than simply as one in a portfolio of livelihood strategies households use to assure their basic needs. The NAP notes that "by engaging more in commercialized agriculture, wealth creation becomes the motivation. Therefore agriculture, as a business, will increasingly serve as a springboard to a better life for Malawi's farming families, providing children in those households with a much broader set of economic opportunities and career choices than their parents had" (Malawi, MoAIWD 2016b, 1). Although improved food and nutrition security for all Malawians rightly remains part of the overall goal for the NAP, the policy seeks to move the dominant orientation of farmers beyond meeting their own subsistence needs and toward more specialized production and a more diversified sector overall. Central to such an agricultural transformation is a significant increase in reliance on markets—both by farming households to secure increased incomes and, in parallel, by all consumers to meet their food needs.

---

12 A new formulation of the development vision for Malawi, the Malawi 2063 document, was launched in January 2021 as this book was being finalized. The goal is "to propel the country towards achieving economic independence, inclusive wealth creation, self-reliance and a high quality of life for all its citizens" by 2063 (Malawi, NPC 2020, 1). The vision is anchored on the three pillars of agricultural productivity and commercialization; industrialization; and urbanization.

However, the sectoral growth proposed in the NAP is not situated within or justified as part of a broader process of economic structural transformation that will see workers in Malawi increasingly move out of agriculture and into more remunerative activities in manufacturing or service provision. For the most part, Malawi's agricultural policy does not provide guidance for building a much more diversified economy that will assure through strong markets the food needs of all, whether working in agriculture or, increasingly, in other sectors of the economy.

The National Agricultural Investment Plan (NAIP) for 2017/18 to 2022/23 (Malawi, MoAIWD 2018) is the strategy to achieve the objectives of the NAP. "Food and Nutrition Security" is the title of one of its 16 intervention areas. In the NAIP's introductory chapter, a useful discussion is provided on how the plan is to consider food security in a significantly more holistic manner than did earlier strategies—in particular, the strategy seeks to go well beyond maize self-sufficiency. However, how to translate this change in conceptual orientation around food security into a change in operational focus for the sector is not spelled out. Under the plan, food production remains the means by which the food security of Malawi will be achieved. The five-year budget for the food and nutrition security intervention area is $209 million, but more than 95 percent of these funds are to be directed to establishing 1 million integrated household farming gardens annually. Although further details on what is involved in this food security-focused effort are not provided, it appears to be exclusively a production-oriented activity. The strategy does not provide any guidance on how the agriculture sector can play an expanded role beyond production in assuring food security through improved incomes and stronger agricultural markets, and more broadly contribute to a structural transformation of the economy away from agriculture.

The NAP and NAIP, if implemented effectively, will entail important reforms to the content of agricultural and food security policy and strategies in Malawi. A review of public expenditure patterns within the agriculture sector in recent years confirms that shorter-term food security concerns addressed through production-focused programs—particularly input subsidy programs—have dominated public sector activities in agriculture (Box 3.2). Agricultural research, agricultural extension, strategic grain reserves, and investments in agricultural market systems are all important factors in accelerating agricultural productivity growth, expanding use of reliable agricultural markets, and in this way, building resilience into Malawi's food systems. However, we see that the concentrated allocation of public expenditure toward input subsidies has been at the expense of these areas in

BOX 3.2 Recent and planned public expenditure patterns in Malawi's agriculture sector

The Monitoring and Analysing Food and Agricultural Policies project of the Food and Agriculture Organization of the United Nations examined public expenditure in support of food and agriculture in Malawi over the period 2006 to 2013 to establish the effectiveness of these expenditures (FAO 2015a, 2015b). Of all public expenditures in Malawi over this period, 17 percent were in support of food and agriculture. Input subsidies under FISP made up almost half of this amount. In contrast to FISP costs, however, expenditures on agricultural support services were much lower. Agricultural research accounted for 2.5 percent of total expenditures on food and agriculture. More troubling, the analysis estimated expenditures on agricultural extension to be only 1.7 percent. The costs of maintaining strategic food stocks amounted to 1.5 percent of public expenditures on food and agriculture. Similarly, relatively low levels of public investment were made in marketing support activities. (Most of the balance of expenditures in support of food and agriculture over this period were on rural roads, with somewhat less on irrigation.)

The five-year, $3.2 billion budget proposed under the NAIP provides insights on whether development priorities within the sector might be changing. This budget reflects total resources considered necessary to implement the plan, not simply government resources. Overall, the budget reflects the new policy stance for agricultural development in Malawi detailed in the NAP. Although food security continues to be seen as best addressed through increased household production, and subsides for agricultural inputs are a prominent component of the plan, other priorities under the investment plan are aligned more closely to a vision of a transformed, expanding, and profitable agriculture sector. More than 60 percent of the NAIP budget is allocated to capital expenditures rather than recurrent costs—these include rural road construction and maintenance (the largest expense in the budget), irrigation scheme construction,

which sustained public investment is needed. It remains to be seen whether the changes in sectoral priorities and in addressing food insecurity laid out in the NAP and NAIP will change public investment patterns in coming years.

Food security is not a sufficient objective on its own to guide agricultural development planning in Malawi to meet the longer-term needs of the economy or its people. Moreover, food security will not be assured using agricultural strategies alone. These are among the chief arguments made in this book. However, it should be expected that agriculture and food security will remain linked for some time to come. First, threats to Malawi's food security demand attention in the short term both because of the danger posed some investments in market structures, the construction of crop storage facilities, and the annual establishment of 1 million integrated household farming gardens. Significantly increased resources are to go to agricultural extension and research, with $400 million—more than 13 percent of the total budget—allocated over the five years, split 75:25 between extension and research.

A shift in development priorities and in expectations about the government's role in how they are to be achieved also can be seen in what is missing from the NAIP budget. For example, whereas access to finance is critical for commercialization of agricultural production and processing, the government does not propose in the NAIP to insert itself into financial markets. Rather, its role will be one of facilitation and risk sharing with financial firms, as well as strengthening the financial literacy and management skills of farmers and processors. Similarly, the government does not specify in detail how it will act to foster agribusiness development. Rather, the resources identified for this intervention area are those that private-sector firms have signaled they will invest under Malawi's country cooperation framework with the New Alliance for Food Security and Nutrition, which seeks to encourage greater private-sector investment in agricultural development (NAFSN 2013). Whether these financial resources will materialize is unclear. Nonetheless, the government states that its principal role in expanding agribusiness under the NAIP will be to establish a business-enabling environment with "predictable policies, supporting legislation, and infrastructure and support services" (Malawi, MoAIWD 2018, 55).

Overall, the NAIP indicates some encouraging changes in how the public sector engages in agricultural development in Malawi. However, its vision for how food security might reliably be attained for all primarily reflects the long-standing and insufficient focus on own production as the principal means by which Malawians access the food they require.

Source: FAO 2015a, 2015b; Malawi, MoAIWD 2018.

to the welfare of individuals across the country and for political economy reasons. The public investment needs associated with longer-term agendas of sectoral or broader economic transformation receive lower priority than do immediate risks to the food security of the country and its vulnerable citizens (Timmer 2015). Food crises anywhere in Malawi have often resulted in the suspension of budgets and agricultural development plans while the crisis is managed.

However, equally important in accounting for this food security focus in the allocation of public resources for agriculture is *who* is involved in implementing agriculture-sector development and support activities. Agricultural experts primarily staff the Ministry of Agriculture. The ministry is not staffed by economic policy experts who could keep attention focused on the needed longer-term economywide changes to which the sector can contribute, nor by welfare and social safety net experts who might suggest other mechanisms than enhancing subsistence production levels to meet the food needs of the most vulnerable households. The political choice to focus on longer-term public investments to bring about a structural change in the Malawian agriculture sector is unlikely to be made without a clearly articulated development agenda. Such an agenda should seek to move the sector away from production for food self-sufficiency at both household and national levels, and toward a market-centered, more concentrated and specialized sector that, nevertheless, will reliably serve the food needs of increasing numbers of Malawians working outside of the sector. This development vision would need to be endorsed by the political leaders of Malawi, who would hold to account those responsible for implementing it.

Chapter 4

AGRICULTURAL MARKETS IN MALAWI

The thin and volatile agricultural markets of Malawi are the focus of this chapter. The conceptual understanding here is that both the dominant subsistence orientation of smallholder farming households and the weak agricultural markets result from two factors: (1) farmers' uncertainty as to whether they will find buyers who will offer them a profitable price for their food crops and (2) consumers' uncertainty that they can reliably find the food they require in the market at a price they can afford. These risks depress traded volumes, undermine market development, and reinforce the subsistence orientation of farming households.

Several factors that drive these market conditions are discussed here:

- Price shocks due to both poor and bumper national harvests
- Unpredictable government restrictions on trade in maize, which impede the access of surplus maize producers to more competitive regional markets and of consumers to greater supplies from Malawi's neighbors
- Often ineffective, but still disruptive, interventions in domestic maize markets by the government through the agricultural marketing parastatal, ADMARC
- Humanitarian responses to large-scale food crises, particularly involving food aid

In consequence of these factors, maize prices in Malawi have been volatile both within the annual cropping cycle and between cycles over the past decade or so, increasing market risks for potential commercial maize producers and for maize traders, inhibiting consumers from more readily relying on the market for the maize they require, and reinforcing the subsistence orientation of any farming they do. However, Malawian households are increasing their reliance on the market for their food, particularly in the lean season before the next harvest and particularly poorer households. There are few other options once a household's food stocks from its own production are exhausted.

Although much more easily stated than achieved, greater predictability in prices will reduce risks for both maize producers and consumers in Malawi, reduce incentives for farm households to primarily focus on low-output subsistence production, and enable many more Malawians to confidently pursue nonfarm employment.

Although farmers in Malawi have important agroecological advantages that they can exploit to produce a wide range of crops for their households' own consumption, the range of crops that they can produce and sell to earn sufficient income to meet other household needs and to increase their assets is considerably smaller. Agriculture in Malawi is dominated in terms of land, labor, and technology used by the low-input, low-output, and low-capital production practices of smallholder farmers. Most rural households have access to land for food crop production. Moreover, about a third of urban households, even if primarily employed in other sectors of the economy, produce some of their own food either through peri-urban farming or by organizing food crop production in their rural areas of origin. Although a growing share of the population seasonally relies on domestic markets for meeting a significant portion of its basic food needs, most farm production is focused on food production for own consumption, with relatively small shares of the total production of food crops being sold (Table 4.1).

The high value that farm households place on self-sufficient food production results in relatively little specialization in crop production among smallholders. In general, the quality of local market performance, particularly

TABLE 4.1 Households that engage in any crop production reporting production of and sales of selected crops, both rainfed and irrigated, in 2015/16, percentages

Crop	Produce crop	Of those producing, who reported any sales	Of those selling, mean share of harvest sold
Maize	95.1	16.0	34.6
Local varieties	*45.0*	*11.4*	*31.2*
Hybrid, recycled hybrids, or improved open-pollinated varieties	*52.3*	*18.7*	*36.7*
Rice	4.3	67.0	57.5
Groundnut	15.5	57.2	58.4
Soyabean	10.2	82.2	76.8
Pigeonpea	22.4	35.1	59.0
Bean or cowpea	14.1	29.9	66.8
Tobacco	6.4	91.3	92.5
Cotton	1.6	91.4	89.1
Sunflower	1.1	76.9	83.9

Source: Author's analysis of 2016/17 Malawi Integrated Household Survey 4 (Malawi, NSO 2017).
Note: Observations: 9,157 households. Weighted analysis.

during periods of economic strain, such as following a poor harvest, is insufficient to eliminate the risk that a producer who brings product A to the market in order to obtain product B from the proceeds of the sale of product A will not find buyers for product A offering a price that would allow the purchase of needed quantities of product B. Given this risk, most farmers will diversify their production as best they can to be self-sufficient in all commodities that they can produce.

Thin Agricultural Markets

The prevalence of production for own consumption largely reflects the persistently thin markets for food crops in Malawi, and also contributes to this market weakness. Farmers are uncertain as to whether they will find buyers who will offer them a profitable price for their food crops, so any surplus production for most farmers primarily is generated as a safety measure against the food security risks of a bad cropping season, rather than as a commercial decision (Dorward et al. 2009). The small transactions they make with their limited surplus food crop production contribute to this pattern of thin food crop markets. An important consequence of the relatively limited supplies of food to domestic markets is that small variations in supply can result in abrupt and significant price changes (Dorward and Kydd 2004).

When considering whether or not to market agricultural produce—especially food crops—farmers contend with a number of factors that keep transaction costs high, including poor roads and, consequently, expensive transport, and limited information on market opportunities, among others. When spread across the low trading volumes that characterize rural market activity, these high transaction costs and uncertainty about current market conditions mean that both buyers and sellers face the risk that a desired market transaction might fail altogether. These risks depress traded volumes, undermine market development, and reinforce the subsistence orientation of the production decisions of farming households.

Maize price volatility

Figure 4.1 plots average monthly maize prices over the period 2009 to 2018 for nine of the larger maize markets across the country, three in each of the three regions of Malawi. Several features of maize price patterns over time and space in Malawi can be seen in the graph.

FIGURE 4.1 Monthly average maize prices in selected markets in Malawi, 2009 to 2018

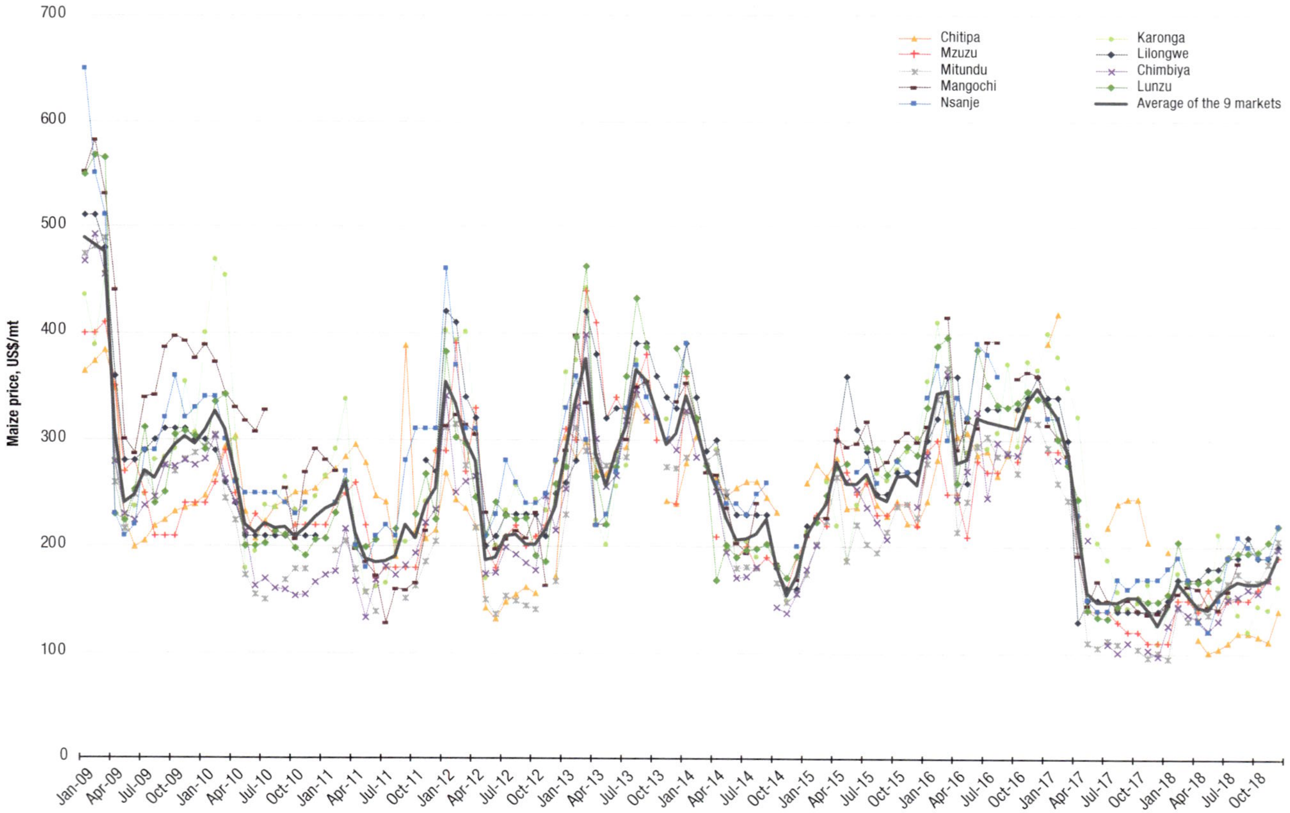

Source: Author's calculations based on prices reported by the Agricultural Market Information Service of the Ministry of Agriculture, Irrigation, and Water Development of Malawi; the International Food Policy Research Institute's Lilongwe monthly maize price bulletin; and the Global Information and Early Warning System on Food and Agriculture, managed by the Food and Agriculture Organization of the United Nations.

- In most years, as many households deplete the maize stored from their last harvest, they will increasingly rely on the market for the maize they consume. In response to this demand, prices rise in the months before the next harvest, peaking around March, when the maize harvest begins in southern Malawi. In other words, there is an underlying seasonality in annual maize price patterns centered around the harvest of rainfed maize. However, this seasonal pattern is not seen consistently across the 10-year period. For example, the figure shows unexpected price spikes soon after harvest in 2013 and 2014 that then moderated for a few months before rising again. Also, the relative change in prices between the immediate postharvest period (May, with the lowest expected prices) and the following preharvest period (February, with the highest) shows significant variation—in the second half of the period examined in Figure 4.1, several years show this price change to have been flat to negative.
- Both the level of annual maize production and the response of authorities and their development partners to any shortfalls in production affect the annual maize price patterns. For example, the 2007/08 maize harvest was poor (see Figure 2.2). Consequently, prices rose significantly in the months before the good 2008/09 harvest and then fell sharply as that harvest came in. The 2015/16 harvest was similarly poor, particularly in southern Malawi. However, in contrast to 2008/09, a large response to the emerging food insecurity was mounted by the government and its partners starting in mid-2016. In consequence, what would have been a sharp rise in prices in the months before the 2016/17 harvest, paralleling the price pattern of 2008/09, was blunted by the importation of significant amounts of maize through 2016 and into 2017 as part of the response. Prices then fell sharply with the good 2016/17 harvest.
- Variability in prices can be seen across the country at any point in time, suggesting that maize markets in Malawi are not sufficiently integrated for price adjustments to occur immediately across the country. However, we also do not see patterns of consistently low or high prices for specific markets, suggesting that price adjustments nonetheless occur reasonably quickly.[1]

---

1 Myers (2013), using weekly maize price data from 2001 to 2008 for 10 markets in Malawi, estimated price adjustment half-lives of between 0.6 and 2.2 weeks, depending on the market pair. (The half-life is a measure of the speed of price convergence between two markets—the time period after a price change in the first market occurs until prices in a second market have adjusted by half the size of the change in price in the first market.) An earlier study by Chirwa (2000), using monthly prices from 9 markets over the period 1989 to 1998, found longer half-lives, of between 5 weeks and 6 months. Both studies concluded that maize markets in Malawi are relatively well integrated, with Myers (2013) suggesting that the level of integration is increasing over time.

A range of policy decisions underlie these maize price patterns, even if they are not their primary drivers. Export and import policies affect the level of the peaks and valleys of the maize price series. With open trade in maize grain, this variability will be bounded by the import and export parity prices for maize. However, with restrictions on trade, maize production shortfalls will result in higher price peaks as a result of the maize supply not being augmented by trade, while maize production gluts will result in lower, less remunerative prices for producers. Similarly, any uncertainty about government policy around domestic maize marketing will result in commercial maize producers focusing their production on other crops and traders being more hesitant to build their stocks of maize for later sale. Both reactions to policy uncertainty exacerbate shortfalls in maize supply in the months before the next harvest, pushing prices higher. Decisions around public investments also affect these price patterns. Improved rural transportation and communication infrastructure can enhance market information flows and efficiency in moving maize from areas of surplus to areas of deficit, further improving integration across the market system. Such investments will result in greater convergence in maize prices across Malawi.

Malawi has had some of the most volatile maize prices in eastern and southern Africa. Using unconditional volatility in monthly prices as a measure, Figure 4.2 shows trends in maize price volatility for specific marketplaces in Malawi, the region, and the world.[2] Over the period 2009 to 2013, the volatility measures for the nine markets in Malawi considered in Figure 4.1 and Figure 4.2 were greater than those of any of the other markets considered. However, in the more recent five-year period from 2014 to 2018, maize price volatility in these Malawi markets has fallen somewhat, whereas it has increased in several markets in Tanzania and Mozambique. Consequently, the high volatility of prices in Malawi's maize markets no longer stands out as prominently within the southern Africa region as it did in earlier years.

Using the average of monthly maize prices across selected markets in Malawi, Mozambique, Tanzania, South Africa, and Zambia, as well as in

2 The unconditional volatility measure used here is the standard deviation of the month-to-month change in the natural logarithm of maize prices over the periods considered (Minot 2014). This measure of price variability is better than the more straightforward coefficient of variation, the value of which is dependent in part on the length of the price series examined. In contrast, the unconditional volatility measure is based only on the observed variation in monthly prices.

FIGURE 4.2 Average unconditional volatility in monthly maize prices in selected markets in Malawi, the region, and the world for 2009–2013 and 2014–2018

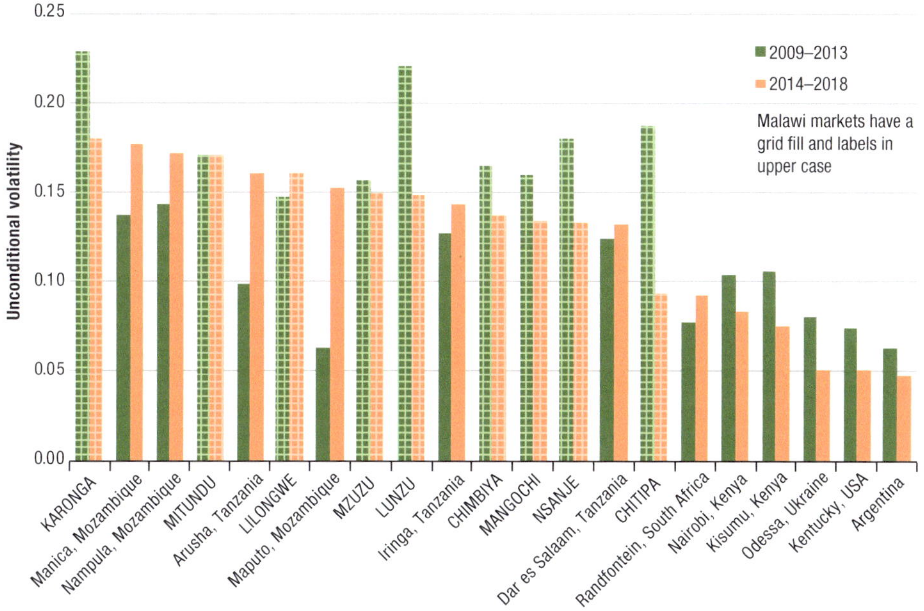

Source: Author's calculations based on prices reported by the Agricultural Market Information Service of the Ministry of Agriculture, Irrigation, and Water Development of Malawi; the International Food Policy Research Institute's Lilongwe monthly maize price bulletin; and the Global Information and Early Warning System on Food and Agriculture, managed by the Food and Agriculture Organization of the United Nations.

Note: Markets are ordered in descending order based on their unconditional volatility in the 2014 to 2018 period.

several non-African markets, Figure 4.3 shows how variability in maize prices changed over the period 2009 to 2013 in these countries. Malawi experienced periods of significant price volatility in 2009 and again in 2012—the latter period principally linked to the sharp devaluation of the Malawian kwacha, by about one-third, in May 2012. However, thereafter, price volatility has moderated. In contrast, both Mozambique, with a price spike in February 2016 followed by a sharp price drop, and Tanzania, with a spike in April 2017, have seen substantially greater volatility in maize markets than

FIGURE 4.3 Moving 13-month average, unconditional volatility of average monthly maize prices in selected countries of southern Africa and the world, 2009 to 2018

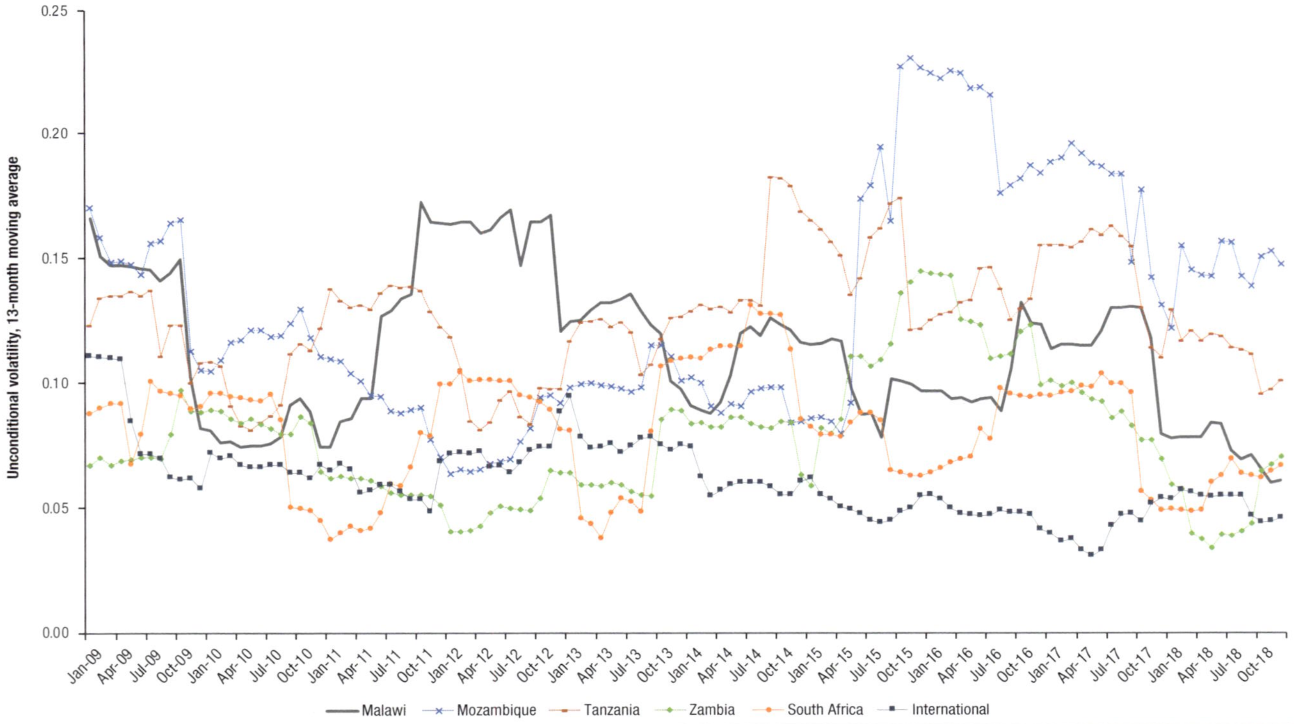

Source: Author's calculations based on prices reported by the Agricultural Market Information Service of the Ministry of Agriculture, Irrigation, and Water Development of Malawi; the International Food Policy Research Institute's Lilongwe monthly maize price bulletin; and the Global Information and Early Warning System on Food and Agriculture (GIEWS), managed by the Food and Agriculture Organization of the United Nations.

Note: The markets from which prices are included in each average national price are as follows: Malawi—Chitipa, Karonga, Mzuzu, Lilongwe, Mitundu, Chimbiya, Mangochi, Lunzu, and Nsanje; Mozambique—Manica, Maputo, and Nampula; Tanzania—Arusha, Dar es Salaam, and Iringa; Zambia—unspecified set of markets from which an average price is reported monthly to GIEWS; South Africa—Randfontein (Johannesburg); and International—Argentina (up-river ports free on board [f.o.b.] price), United States (Kentucky f.o.b.), and Ukraine (Odessa f.o.b.).

were seen in Malawi markets in the second half of the period reviewed.[3] In contrast, international markets and the main reference market for maize, South Africa, show relatively low volatility. This is to be expected given the strong integration of these markets into regional and global maize trade. The pattern of maize price volatility in Zambia over this period is generally lower than in Malawi. That the two patterns do not mimic each other more closely is surprising—Zambia is not very different from Malawi in terms of the risks to maize production that the country's farmers face, the policy stance of its government around maize commercialization and trade, or the role of the commodity in national food security.

No single factor accounts for the patterns of price volatility seen in Figure 4.3 in Malawi or in the other countries. In any of the countries examined, several factors are likely involved in periods of heightened maize price volatility, including shortfalls or gluts in seasonal production levels; macroeconomic factors, particularly sharp currency devaluations; changing policy stances on openness to trade in staple foods; and how a country's leaders balance policy goals for agricultural commercialization against those for food security. The effective response of the government and its partners to broad maize production shortfalls, especially following the poor 2015/16 harvest, has certainly moderated maize price swings in Malawi in recent years, particularly in the Southern region, which has been most affected by poor harvests.

Subsistence production as a response to unreliable food markets

When faced with uncertain markets, producers tend to devote more land to low-value food staples and less land to potentially higher-value, nonstaple crops. Fafchamps (1992) concludes that unpredictable prices limit farmers' incentives to produce cash crops for the market and drive them to be more subsistence oriented than larger farmers, whose relatively lower share of staple food in their consumption expenditure enables them to tolerate more price risk. Alwang and Siegel (1999) find that lack of confidence in markets causes

---

3 Increased maize price volatility in Mozambique is likely linked to the debt crisis that emerged in early 2016. This had an adverse impact on several macroeconomic indicators. In 2016 the country experienced its highest rate of inflation in many years and a devaluation of Mozambique's currency, the metical, by 45 percent in US dollar terms. For Tanzania, the recent increases in maize price volatility were due to high maize prices in early 2017 that resulted from poor harvests in bimodal rainfall areas in the second cropping season of 2016 (FEWS NET 2017). These were followed by very low prices later in the year after a good national harvest in the main cropping season. This price pattern was repeated in 2018 with high prices early in the year that then fell sharply.

farmers to use a large part of their land to produce low-value food staples. As a result, farmers are constrained in their ability to purchase inputs or other productivity-enhancing technologies. In a review of evidence on smallholder market participation in staple food markets in eastern and southern Africa, Barrett (2008) concludes that high commerce costs limit household-level market access, price transmission, and trader competition, leading to more volatile markets. These volatile markets, in turn, limit farm-level incentives to generate surpluses for the market through increased productivity, and thus curtail the potential of the agriculture sector to contribute to economic growth and poverty reduction.

Unpredictability in the seasonal pattern of maize prices also has an adverse impact on maize traders, particularly those who store maize obtained just after harvest in order to sell it later, when prices can be expected to peak before the following maize harvest. (The same adverse effects also face producers who store any surplus for later sale to profit from expected higher prices.) Engaging in storage is profitable only when the seasonal rise in maize prices is sufficient to generate a positive return on the investment the trader made in storage facilities as well as cover the costs of managing the grain while it is in storage (Chapoto and Jayne 2009).[4] However, if the risk that prices will not rise as anticipated is increasing, storage of grain for later sale will not be a financially viable business strategy for traders to pursue—those storing maize will often realize negative returns (Cardell and Michelson 2020). Increasing unpredictability in seasonal maize price patterns in Malawi in recent years could be due to a combination of rising incidence of production shocks, both positive (gluts) and negative (deficits); increased discretionary interventions by government that affect maize marketing and trade patterns; and more frequent humanitarian responses to food crises that involve significant food aid. Any breakdown in the predictability of seasonal patterns in maize prices and consequent reduction in the storage activities of maize traders will be detrimental to Malawian consumers who obtain some of their maize from the market. Without storage, there will be less maize available in the market in the months before the next harvest. A stable, predictable pattern of prices over the year is characteristic of a strong and reliable market.

Unpredictable prices also hurt Malawi's consumers of maize, especially when one considers seasonal price movements. Analysis of the average maize

4 Myers (2013), using weekly maize price data from the period 2001 to 2008 for 10 markets in Malawi, found that storage would be an efficient way to manage temporal price movements in 8 of the markets over this earlier period.

price series, presented in Figure 4.1, shows that the average monthly price of maize in Malawi over the period 2010 to 2019 increased by one-third on average between the period just after harvest in May and the height of the lean season in February, when many households have exhausted their own stocks and look to the market for maize. However, there is considerable variability in the magnitude of these annual price movements—average maize prices increased by 80 percent in 2011/12 and again in 2012/13, whereas in 2014/15 and in 2017/18, prices essentially did not change between May and the following February. These potentially large but inconsistent price spreads affect how households plan to meet their food needs throughout the year and how they cope with any shortfalls in their own production of maize and other foods, especially in years in which maize and other food prices are on the high side. Such uncertainty in maize prices also makes it perilous for households to diversify from maize production into other, more profitable crops or nonfarm activities, because they depend so heavily on maize to meet their dietary needs (Dorward and Kydd 2004). Households cannot be confident, in the face of potentially high maize prices, that shifting their production away from maize to nonstaple crops will be sufficiently remunerative to assure them the income needed to buy the maize they require.

Although the large seasonal price swings make the market a somewhat unreliable means for gaining access to food, many households rely on the market to meet their food needs. In 2016/17, survey data show that the average share, by weight, of per capita maize consumption for all households that was obtained from the market doubled in the lean season, when maize prices were higher, compared with the harvest period, between April and June (Figure 4.4).

Moreover, reliance on the market to obtain maize is increasing among Malawian households, in both rural and urban settings (Figure 4.5). Purchase of maize for household consumption increased slightly between 2004/05 and 2010/11. But this reliance on markets increased sharply between 2010/11 and 2016/17. We also see in rural areas considerably more reliance on gifts of maize in 2016/17 than was the case in the two earlier survey years, both of which followed reasonably good maize harvests. The poor harvest in 2015/16, and the consequent food insecurity response mounted by the government and its development partners, is part of the explanation for this change in the sources of maize for household consumption reported in the 2016/17 household survey year. However, qualitative studies by FEWS NET of the sources of food consumed by Malawian households suggest a similar trend over the

FIGURE 4.4 Pattern in sources of maize consumed by Malawian households, by month, May 2016 to April 2017, percentage of total maize consumed

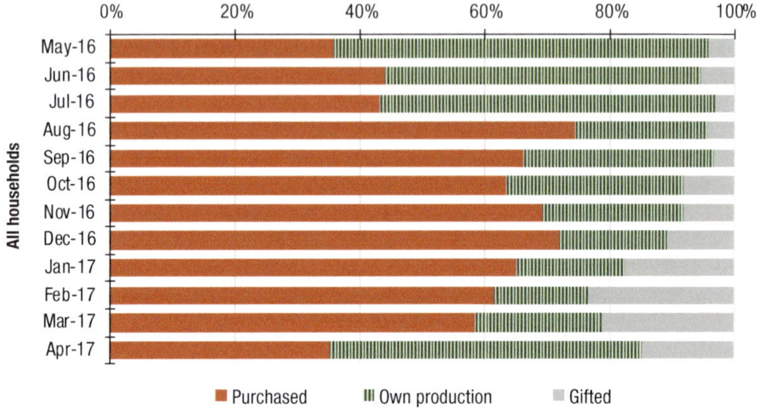

Source: Author's weighted analysis of 2016/17 Malawi Integrated Household Survey 4 (Malawi, NSO 2017), based on the sample of 11,697 households interviewed over the 12-month period who reported consuming maize in the week prior to being interviewed.

FIGURE 4.5 Source of maize consumed in 2004/05, 2010/11, and 2016/17, by rural and urban households and by household welfare quintile in 2016/17, percentage of total maize consumed

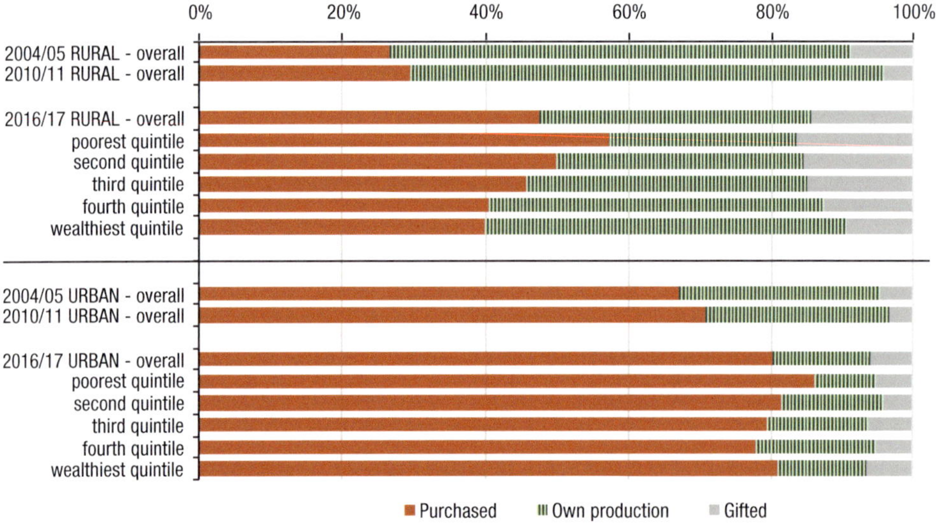

Source: Author's weighted analysis of data from subsample of survey households interviewed over a 12-month period who reported consuming maize in the week prior to being interviewed—2004/05 (Malawi Integrated Household Survey 2): 10,004 households interviewed between April 2004 and March 2005; 2010/11 (Malawi Integrated Household Survey 3): 11,384 households, April 2010 to March 2011; 2016/17 (Malawi Integrated Household Survey 4): 11,697 households, May 2016 to April 2017 (Malawi, NSO and World Bank 2007a; Malawi, NSO 2012, 2017).

past 15 years of decreasing reliance on own production and increased reliance on market purchase (FEWS NET 2018).

In addition to increased reliance on the market across all income groups, poorer households are more likely to purchase their maize than are wealthier households, particularly in rural communities. Moreover, Mussa (2015) has shown that poorer households in Malawi generally pay more for maize than do other households in both rural and urban settings, attributing this primarily to the poor generally purchasing maize in small quantities at a time due to cash constraints. Consequently, they do not benefit from quantity discounts on the maize price. The implications of maize price volatility for the food security and well-being of Malawi's poorest households are more pronounced than for their wealthier neighbors.

Government response to maize price spikes

When maize price spikes have occurred in Malawi outside of the normal seasonal pattern of price variation, government interventions in the market to manage such perceived food crises have often exacerbated such spikes. Although maize price volatility generally originates in production challenges that constrain supply, it also is associated with discretionary government interventions in the market, including ill-timed procurement or stock releases; mixed signals from government concerning intended procurements or price controls; uncertainty concerning the imposition or lifting of an export ban on maize; and unreliable estimates of maize production. Both traders and producers of maize face considerably greater risk in their commercial enterprises around maize than would be the case if the government adopted a predictable, rules-based approach to how it intervenes in agricultural markets.

Ellis and Manda (2012) provide an account of three episodes of maize price spikes in Malawi between 2001 and 2009, when prices rose between 200 and 400 percent. They describe a recurring pattern that starts with the government, primarily for political reasons, being slow to recognize a maize production shortfall, despite atypical price increases in maize markets. Thereafter, as prices continue rising, the government, through ADMARC and the National Food Reserve Agency, scrambles to obtain sufficient stocks domestically or regionally, its actions in the market raising maize prices further. At the same time, the government bans exports and presses grain traders to release the typically nonexistent grain stockpiles that they are believed to have hoarded. The government then institutes a ban on private maize trade, eliminating any incentives for traders to engage in efforts to supply additional maize to Malawi, including through cross-border trade, even though

ADMARC, the sole remaining supplier, has limited stocks of grain. Prices spike further, only starting to fall with signs of a good harvest in the following cropping season.

Between 2011 and 2016, Malawi experienced three further episodes of rapidly rising maize prices (as well as a period in the second half of 2014 of abnormally low maize prices). Although maize prices over this period, in real terms, did not reach the levels of the earlier spikes documented by Ellis and Manda (2012), many of the generally ineffectual actions that the government took in an attempt to mitigate the earlier spikes were repeated over this period.

In 2016 Malawi again faced significant food insecurity following a 25 percent drop in production from normal levels in the national maize harvest for 2015/16 due to poor rains in southern Malawi caused by a strong La Niña phase in the El Niño–Southern Oscillation climate pattern. The scale of the crisis, with an estimated 6.7 million Malawians vulnerable to food insecurity in the postharvest period, was such that the government and its development partners developed a joint response plan to address the food needs of this vulnerable population—the 2016/17 Food Insecurity Response Programme. More than $235 million was spent in the form of cash, food, and vouchers to supply immediate food aid and to contribute to the restoration of livelihoods in affected areas (Babu et al. 2018). The response program was generally quite effective— there was no spike in maize prices in late 2016 and early 2017 (see Figure 4.1) and maize prices dropped sharply once the good 2016/17 harvest began. Moreover, no evidence was found that the provision of food aid distorted local market prices.

The 2016/17 Food Insecurity Response Programme demonstrated that broad food insecurity crises in Malawi can be managed so that significant short-term maize price volatility does not undermine household food security, erode household welfare, or hamper the development of commercial agriculture. Price spikes and collapses do not need to be recurrent features of maize markets in the country. However, the costs of managing supplies of maize to meet demand through a humanitarian response is high, and as the 2016/17 activity showed, the capacity of the government to intervene in the provision of staple food and to ensure that all households have access to the food they require was limited. Given the urgency of the crisis, the UN's World Food Programme took overall responsibility for the response to food insecurity down to local levels, with most government leaders and agencies at district and more local levels largely left out of the implementation process (Babu et al. 2018). The government's maize procurement efforts for the response were not well managed, with both delays and inflated costs. Overall, the 2016/17 experience demonstrated that government did not yet have the institutional capacity to respond to and effectively handle large-scale food crises without significant assistance from development partners.

Although international assistance will remain an important element in responding to future large food insecurity crises in Malawi, the limited ability of government to respond to maize shortages requires an increased and strengthened role for market mechanisms to respond efficiently to such shortfalls in domestic maize supply. That Malawi finds itself on the verge of significant food insecurity following each poor cropping season is due as much to deficient domestic food market mechanisms as it is to the risks inherent in reliance on rainfed crop production. The current structure and operations of agricultural markets in Malawi are inadequate to reliably provide for the food needs of vulnerable households. Although building increased capacity within government to respond to food crises is important, a more sustainable solution to recurrent widespread food insecurity in Malawi would be centered on efficient markets that provide sufficiently profitable prices to producers who sell their maizeand which in turn supply that maize at affordable prices to households across the country that are in need.

Government's Direct Engagement in Food Crop Marketing

For the agriculture sector to be an active part of a more open, market-based economy, a rules-based approach to government intervention in the maize market needs to be put in place. The way the government of Malawi has historically engaged in supporting commercial activities has not been through supporting private firms to competitively engage in production, marketing, processing, and retailing. Rather, the government generally has addressed the problem of inadequate marketing services for the crops produced by smallholder farmers or for the food crops required by Malawian households, particularly maize, by monopolizing trade through parastatal agricultural marketing organizations or by maintaining close oversight of marketing and trade in such crops.

Quite early on, the colonial government tightly regulated the sale of the crops produced by smallholders, instituting crop marketing boards. Richard Kettlewell, the director of agriculture for the colonial government in the decades before independence, later provided this reasoning for the establishment of such marketing boards:

> *If African cultivators were to be converted from subsistence to a cash economy it was felt that they must have confidence that there was a continuing market for their produce at fair and reasonably steady prices, and an efficient organization to buy it.... Government concluded that this aim could not be fulfilled if marketing were left to private enterprise. The fluctuations of price inherent in such a system would bewilder and discourage the inexperienced farmer .... Government control of the marketing of the more important crops produced by African farmers therefore appeared essential. (Kettlewell 1965, 248)*

To a large degree this protective and somewhat patronizing perspective on the role of government in regulating the marketing activities of smallholder farmers in Malawi is maintained today in policy debates around food security and agricultural development. In 1971, the parastatal ADMARC was established to replace the colonial-period marketing boards. Although initially ADMARC was considered significantly more efficient than the parastatal agricultural marketing corporations of Malawi's neighbors (Dorward and Kydd 2004), its effectiveness declined as it increasingly came to be used by government as a vehicle to tax smallholders through low prices for their cash crops, transferring significant resources from commercially oriented smallholder producers to the emerging tobacco-producing small estate sector in the late 1970s and early 1980s (Kydd and Christiansen 1982). ADMARC also played an important role in the food security of the country, exercising its monopsony to purchase smallholder-produced tobacco, groundnut, and cotton at prices substantially below export parity, and using some of the revenue obtained to subsidize both producer and consumer maize and rice prices (Harrigan 2001).

The structural adjustment reforms of the 1980s that Malawi negotiated with its donors resulted in reforms to ADMARC to ensure more remunerative prices for smallholder producers. Its formal monopoly as the buyer of smallholder produce, though never complete, was eliminated in 1987. Liberalization of agricultural markets for smallholders continued through the 1990s.

Only for maize has ADMARC continued to be involved in price stabilization efforts—the government sets a price band that the parastatal is expected to defend through its maize purchases and sales strategies. However, ADMARC usually does not receive sufficient financial resources in a timely manner from the government to effectively stabilize maize prices across the country from year to year. Consequently, the amount of maize it handles in a year is usually significantly less than would be required to keep prices within the established price band (Baulch and Botha 2020). Nonetheless, the extensive network of local ADMARC depots across the country (Figure 4.6), including in more remote areas, provides the government with the physical means to distribute food in times of shortage. ADMARC today remains one of the central mechanisms that government has at its disposal to intervene in agricultural markets in Malawi, and it is seen by most Malawians as an important, if not always effective, mechanism for assuring their food security (Chilowa 1998; Chinsinga 2015).

The government is generally cautious, even distrustful, of the ability of private traders to be active in Malawi's food and agriculture markets under all conditions of relative supply and demand. The low capitalization of most traders, their limited access to financing, and the relatively high transaction costs associated with agricultural marketing in Malawi due to poor transport infrastructure and inadequate information and communication services all limit the capacity of private traders to move sufficient maize and other crops from sources of supply to consumers (Ochieng, Botha, and Baulch 2019). Because ADMARC, as a parastatal, has the resources to move maize stocks to its depots across the country, even if those stocks are quite small, its maize sales can serve to temper local maize price spikes.

However, ADMARC does not operate efficiently, limiting its ability to effectively and reliably moderate movements in maize grain prices. Although it is considered a commercial enterprise, it undertakes a range of social functions for the government, most notably related to food security, that by their nature often are not profitable. ADMARC looks to government to cover the costs of these social functions. However, government generally is slow in providing funds, limiting ADMARC's ability to engage in maize purchases soon after harvest to build the maize stocks that will enable the agency to make maize available to consumers quickly to blunt any price rise. Typically, ADMARC buys maize from producers late, several months after harvest, and often has insufficient grain stocks to meet consumer demand in the lean season, forcing restrictions on the amount consumers can buy. Confusing or late government directives and poor timing of its marketing activities mean ADMARC often sells its maize at a lower price than it was purchased for. The significant losses that ADMARC has repeatedly incurred are the responsibility of the government to cover—for example, a bailout of ADMARC by the government in early 2018 involved an unbudgeted expenditure of 45 billion Malawian kwacha (more than $60 million), equivalent to just under 1 percent of the country's GDP.

FIGURE 4.6 Location of ADMARC depots relative to major roads, 2002

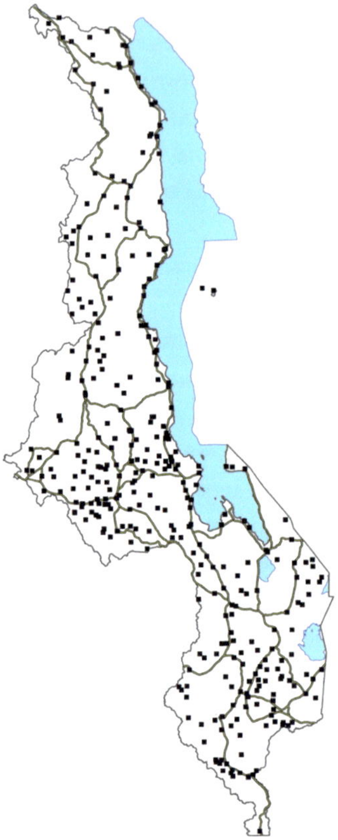

Source: Author's analysis, prepared from a comprehensive list from 2002 of about 360 ADMARC unit markets (permanent retail-level depots).

Note: Although there has been significant closing and reopening of depots since 2002, a 2015 unpublished document prepared by ADMARC for the Public Sector Reform Commission noted that the corporation still operated more than 300 unit markets and 9 storage depots, with total storage capacity of more than 400,000 metric tons. The 2015 document also noted that 205 of the unit markets were uneconomical to operate.

Efforts to restructure ADMARC are promoted regularly, particularly by donors, as part of programs to reform agricultural markets in Malawi, but so far with little to show. Such reforms are certainly needed. For example, clearer guidelines on how ADMARC is to serve the policy priorities of government could increase efficiency. However, any reforms to ADMARC will also need to determine how best to provide incentives to the private sector so that many more firms engage in the marketing and trade of food and other agricultural products. When significantly more, and more competitive, agricultural trading firms emerge, the private sector will increasingly be able to undertake—in a more cost-effective manner—the functions that the government now often calls upon ADMARC to perform.

Continuing significant public investments in ADMARC are made at the expense of increased private-sector engagement in maize marketing. Consequently, as part of efforts to improve food security in Malawi, a reasonable longer-term policy objective should be to eliminate ADMARC. The generally ineffective efforts of ADMARC to defend the government-set maize price band add uncertainty to the market, adversely affecting all concerned—maize producers, traders, and consumers. In terms of market development, the funding and policy decisions through which the government supports ADMARC pose significant risks for private firms, both Malawian and regional, that might engage in maize trading in the country. These firms potentially could be efficient competitors to ADMARC, managing at lower risk and at lower public cost many of the functions ADMARC ostensibly has been set up to undertake. So long as the government's discretionary support to ADMARC distorts agricultural markets in Malawi and makes them more uncertain in terms of the commercial returns they might offer, many regional agricultural trading firms will choose to operate their maize businesses elsewhere than in Malawi's markets, and domestic firms will be constrained in the scope of their operations. In the medium term, a purely private sector–managed food system in Malawi should serve as a goal for how the food needs of the population are met under normal circumstances.

However, despite the ineffectiveness of government engagement in the food system through ADMARC, it will be some time still before the parastatal's role is substantially reduced. Given its scale and centrality to government actions to promote food security, reforms to ADMARC are needed to make it more effective. At the same time, however, serious efforts must be made to strengthen and expand the role and capabilities of private grain traders and other firms so that ADMARC becomes an unneeded feature in Malawi's food system.

International and Regional Markets

If agriculture is to drive significant economic growth in Malawi, external markets, particularly regional markets in southern and eastern Africa, will need to play a much bigger role in the near future in providing an outlet for increasing production from Malawi's farmers. Until transformations to deepen domestic agriculture and food markets, including defining a proper role for ADMARC in these markets, are well established, the profits Malawian farmers can reliably achieve from focusing solely on domestic consumers will be slim. Malawi's trade has depended on agricultural exports since early in the colonial period—tobacco, tea, and cotton alone made up 83 percent of the value of Malawi's exports in 1960 (Kettlewell 1965) and these three crops still made up 61 percent of exports in the period from 2015 to 2017. In this recent period, overall agricultural exports provided 87 percent of Malawi's exports by value. Clearly export crops are still the basis for most of Malawi's economic engagement with the rest of the world.

However, Malawi could benefit more than it does from its agricultural exports. The Monitoring and Analysing Food and Agricultural Policies project of the FAO has closely assessed the performance of six of Malawi's export crops, including maize, over the period 2005 to 2014, particularly looking at how these export value chains are organized and the policy environment within which they operate (FAO 2015a). Overall, the analysis shows that producers of export crops in Malawi generally have faced policy-based disincentives to production over the period examined. This is even the case with production of maize, which, though benefiting significantly from some policies—most notably FISP—nevertheless is sold within a regulated national output market that often includes a ban on maize exports. These restrictions cap the price maize producers can obtain for their grain.

The low costs of family labor used in smallholder-based production of export crops in Malawi, such as tobacco and cotton, provide Malawi with a significant advantage in production for regional and international markets (Keyser and Tchale 2010). Family labor is implicitly valued at less than the prevailing wage rate for casual labor, keeping production costs low. For other smallholder-produced crops, such as maize, however, the generally low levels of productivity relative to potential productivity make Malawi's producers uncompetitive in international markets against higher-productivity producers. The low productivity in Malawi is partially attributed to the high costs of commercial inputs, particularly of fertilizer, which is imported, and the cash constraints faced by smallholder farmers that restrict their access to available inputs in the absence of government subsidies (Keyser and Tchale 2010).

Efforts around several specific trade-related policy issues have a bearing on whether Malawi's agricultural exports will be competitive within international or regional markets:

- Trade levels are sensitive to the exchange rate for the Malawi kwacha. The foreign exchange controls in place until mid-2012 resulted in a significant loss in the price competitiveness of Malawi's agricultural products, reduced investor interest, and caused distortions within the economy, exacerbating existing macroeconomic instability (Pauw, Dorosh, and Mazunda 2013; FAO 2015a). Removal of the controls sharply reduced the value of the kwacha and led to significant micro and macro adjustment shocks. However, doing so also significantly improved the competitiveness of Malawian producers in export crop markets.
- Malawi is disadvantaged in international trade due to its high costs of transport, both on rural feeder roads to domestic markets and from Malawi to ocean ports or regional market centers.
- Malawi's export performance is tied closely to tobacco, which accounted for just over half of the value of all exports between 2015 and 2017. Diversifying Malawi's agricultural exports away from their heavy reliance on tobacco would mitigate some of the adverse effects on Malawi's economy and its producers of any downward trend in global tobacco markets.
- Information flows within export crop subsectors can be improved. Smallholders are particularly disadvantaged in this regard, having far less information on current market prices and conditions than the buyers of their output.
- Limited competition among buyers in most export value chains of Malawi restricts opportunities for broad economic benefits, particularly for producers. Due in part to the information constraints noted, producers are generally price takers. Moreover, there are opportunities for collusion among output buyers. Efforts are needed to expand the pool of buyers of Malawi's export crops to foster more competitive price-setting processes for those commodities.

Expanding and diversifying the export crop sector in Malawi, including for food crops, will be an important component of agricultural transformation processes in the country (FAO 2015a). Coupling resilient export crop value chains with much thicker domestic markets for food crops would provide the foundation for more specialized producers of food crops or export crops to operate at significantly higher levels of productivity. Broader growth across the economy should result as productive and profitable agricultural producers establish stronger economic linkages with other sectors. Such economic growth would expand the range of employment options for Malawian workers and improve their households' access to the food they need.

Commercial Crop Production and Food Security in the Context of Unreliable Markets and Production Constraints

Price reform was the central focus for agriculture sector reform in the structural adjustment programs prescribed by Malawi's development partners in the 1980s and 1990s. These donors felt it was in the interest of Malawi's farmers to remove policy-related price distortions so that they would have incentives to pursue a broader range of production choices more aligned with their objective comparative advantage. However, many of these price reforms were not fully implemented in practice. Government officials were hesitant to do so in part due to their fear that the policy change would put the food security of the country at greater risk. Lessons from this earlier experience of market reform are worth bearing in mind in formulating development strategies that will expand commercialized smallholder agriculture in Malawi without neglecting the country's food security.

Smallholder farmers in Malawi historically have shown that they respond quickly to higher crop prices with increased output. The immediate response of the colonial government to the 1949 famine was to significantly raise maize producer prices nationwide in 1950. This resulted in a rapid supply response from farmers across the country. Production doubled in 1951 and then almost doubled again in 1952, leading to problems of maize storage and losses for the colonial government as it disposed of its maize stocks on the glutted regional market (Vaughan 1987). A similar pattern was seen in the early 1980s as the government built up its strategic grain reserve stocks following a poor national harvest in 1980 (Harrigan 2001). Improved prices for maize between 1980 and 1984 resulted in "peasant producers ... not only fully [supplying] the domestic market but also a significant component of the food requirements of neighboring countries" (Kydd 1985, 341–342). In the mid-1990s, the national burley tobacco harvest shot up as restrictions on smallholder production were removed.

However, with improved incentives for production of nonfood export crops, burley tobacco, in particular, the challenge facing the government in assuring national food security is that there is limited scope for increased production of export crops without reducing food crop production, at least under current production levels and landholding patterns. Sahn, Arulpragasam, and Merid (1990) provide evidence that much of the high supply response of Malawian smallholders to improved incentives for export crop production is generated by bringing land and labor previously used for other crops into the production of a now relatively higher-priced crop. Given growing land constraints and the continuing challenges that poor farmers face in raising their crop productivity levels using modern technologies, farmers are even more likely now than 25 years ago to be able to significantly increase the supply of a commercially attractive crop only by curtailing production of another that has a relatively lower price. Overall, aggregate agricultural supply across all crops can be expected to change little simply as a result of changes in the relative prices of the crops produced by Malawian farmers (Harrigan 2003).

If smallholders significantly increase production of export crops because prices of those crops become relatively more attractive than those of maize and other food crops, there is a risk that the country's food security will suffer. If the government's strategy for agricultural development involves significant export-oriented production by smallholders, there will remain a continuing need for government support for increasing maize production. This will require both technical mechanisms to increase productivity and price incentives to ensure that farmers commit sufficient farmland to maize, as well as significant opening of Malawi's maize market to imports. In the absence of such production and/or import increases, maize prices will rise with increased export crop production and commercial maize production then will expand, likely at the expense of export crop production by smallholders. Although the higher maize prices will be beneficial for smallholder producers, they will adversely affect most food-insecure households, who will see a rise in the price of their staple food.

These trade-offs illustrate the difficulty of developing a more commercially oriented agriculture sector as part of a broader process of structural transformation of the economy of Malawi, while still ensuring sufficient food production to meet the nation's food needs. These are difficult objectives to achieve jointly. The pattern in maize prices in Malawi seen in Figure 4.1 shows no sign of convergence at a price level that would offer sufficient returns to producers while not exacerbating the food insecurity of the many poor households in Malawi that rely on the market for much of their maize. Propelled primarily by maize production shortfalls and gluts, but often worsened by ill-advised government engagement in domestic markets and in trade, these price variations suggest that agricultural markets in Malawi, and maize markets specifically, are generally unreliable. Greater stability and predictability in prices would reduce market-related risks for both maize producers and consumers in Malawi, reduce incentives for farm households to primarily focus their efforts on low-output subsistence production, and enable many more Malawians to confidently pursue nonfarm employment.

Chapter 5

RECOGNIZING RURAL ECONOMIC DIFFERENTIATION IN MALAWI

This chapter describes an approach to rural economic transformation in Malawi that places commercial smallholder farmers at the center of such efforts. This subset of smallholders is uniquely positioned within rural communities to serve as an engine of economic growth. As their productivity increases, their farm production expands, and their incomes increase, they will demand more of the labor-intensive and generally nonmarketable (outside of the local area) goods and services that their less agriculture-focused neighbors produce. This consumption linkage diffuses many of the economic gains commercial smallholders make from their more productive farming to those other rural households, deepening local markets, accelerating local economic activities, and improving access to food for economically active households in these communities, including the poor.

As the returns they obtain from those nonfarm activities begin surpassing those that they can obtain from their low-productivity farming, many of the households producing goods and services for the local market will expand their activities to serve wider markets, further increasing their income and, of interest here, better ensuring their access to food. Many will transition from being poor, subsistence-oriented households that engage in some farming to become nonfarming households specialized in livelihoods outside of agricultural production.

The first half of this chapter presents an overview of this model of rural economic transformation, with particular attention to the social and economic structure of rural Malawian communities and the opportunities and challenges that structure presents. In the second half, a recent household survey is used to empirically evaluate how realistic would be a rural economic development program centered on commercially oriented smallholder farmers. This assessment shows that it would be challenging, because the number of households in this category is small. Nevertheless, with effective market development and increased agricultural productivity, commercial smallholder farmers as a share of rural households will grow. At the same time, with stronger markets and more reliable access to food through them, other rural households will see many of the barriers and risks to adoption of nonagricultural livelihoods become less constraining.

Given increasing population pressure on a limited land base, the challenge of ensuring the food security of the country primarily through widespread low-productivity, subsistence-oriented farming is likely to intensify. Critical to eliminating hunger and malnutrition, significantly reducing poverty, and fostering broad economic growth will be policies, programs, and public investments that provide incentives for Malawians to improve their well-being and to secure the food they require through economic pursuits outside of subsistence agriculture. In parallel, investments will be needed to improve the productivity of the many Malawians who will continue farming, with their increased surplus production strengthening supplies of maize and other food crops for local and national markets. As was discussed in the last chapter, an element central to such an economic transformation is the improved functioning of domestic markets so that all Malawian households can safely rely on them for much of the food that they consume.

Drawing on conceptual literature and lessons from other countries, this chapter discusses how the government of Malawi can enable many Malawians to achieve their ambitions, whether through agriculture or, increasingly, by engaging in other economic pursuits. This involves a sharper focus on commercially oriented smallholder farming households to increase agricultural productivity and to propel rural economic transformation processes. It is the increased productivity and expanding scale of production of these households that will accelerate economic activity in rural communities across Malawi both in agriculture and more broadly. The success of these farmers will foster increased demand for the economic output of their neighbors, whether in the processing and marketing of local produce; the provision of specialized goods and services, particularly those based on local, generally nontradable resources; or in commerce. However, such a rural economic development strategy must not neglect the food security of other rural households in the short to medium term, even as efforts are made to enable these other households to improve their welfare through livelihoods outside of agricultural production.

A Conceptual Typology of Malawian Households to Guide Strategic Economic Development

The agriculture sector and, indeed, the bulk of the population of Malawi is commonly considered to be a relatively undifferentiated mass of smallholder farming households that are engaged in low-productivity farming of food crops. This assumption should be rejected. Viewing all of those who farm the land as having a similar role in ensuring the food security of their households and communities and as contributing equally to the economic performance of the agriculture sector is misguided. Doing so results in important missed opportunities for promoting longer-term rural economic development and for sustainably improving the welfare of rural communities across the country.

Rather, there is practical significance in differentiating Malawi's population into four groups:

1. A relatively small segment of the rural population made up of nonpoor, commercially oriented smallholder farming households that produce considerably more crop output than they consume within their own households.

2. Poor, subsistence-oriented rural households that engage in some farming while also pursuing a diverse set of generally unskilled, labor-intensive livelihood-earning activities. These households make up the bulk of the rural population.

3. Households that are not economically productive. These include chronically poor households, both in rural areas and urban centers, with insufficient labor, often because of disability, old age, or chronic illness. Also in this category are households that are experiencing a sharp, likely temporary, reduction in their welfare due to a shock to their livelihoods.

4. Households resident in urban centers and rural towns that primarily specialize in economic activities outside of agricultural production. These include households living in rural market centers and supplying goods and services to their neighbors engaged in farming.

The poverty analysis of the IHS4 from 2016/17 allows one to roughly gauge the relative sizes of these groups. That analysis found that 52 percent of rural households in Malawi were poor, having a consumption level below what is required to meet their basic needs (Malawi, NSO and World Bank 2018). Some of the rural nonpoor are not farmers, such as fishing households, so this leaves about 40 percent of rural households that could be commercially oriented smallholder households. (However, whether these nonpoor farming households will be strongly oriented toward producing crops for the market cannot be gauged from the IHS4 poverty analysis.) Most other rural households will be subsistence oriented in the limited farming that they do while also pursuing a range of other livelihood strategies. Assuming that around 50 percent of rural households fall into this group seems reasonable. Depending on the poverty threshold used, chronically poor households unable to engage in any economic activities are unlikely to make up more than 10 percent of rural households.

We now discuss from a conceptual perspective the characteristics of and the most promising development pathways for the two economically productive rural household categories—the commercially oriented smallholder farming households and the other productive households that are more subsistence oriented in their farming activities. Later, a more comprehensive analysis of the IHS4 dataset is used to further explore, specifically for Malawi, the size and characteristics of all four household groups. Due to the weak market participation and continuing strong subsistence orientation of farming households, as well as inconsistencies in the IHS4 data on crop production and crop consumption, the analysis, as will be seen, has its limitations. Nonetheless, insights are gained into how public investments might be made to accelerate economic development in rural communities across Malawi.

Focus rural economic transformation efforts on nonpoor, commercially oriented smallholder farming households

If the agriculture sector is to drive economic development in Malawi's rural areas, there are strong arguments for putting commercial smallholder farming households at the center of such development. As smallholder commercial farming households become more productive, their demand increases for the labor-intensive, generally nontradable goods and services that their neighbors can provide. This consumption linkage diffuses many of the economic gains commercial smallholders make from their more productive farming to those other households, deepening local markets, accelerating growth of local economic activities, and improving access to food for economically active households in these communities, including the poor. In sum, there are significant positive spillover and multiplier effects in rural communities from the economic success of these commercial smallholders.

There is an extensive research literature on linkages between the smallholder farming sector and the rural nonfarm sector in developing countries that goes back at least 50 years. Haggblade, Hazell, and Reardon (2007) provide a detailed overview of several different dimensions of this research. The mechanism for rural economic growth advanced here has been most clearly described by Mellor (2017; also Mellor 2014), who draws on case studies and empirical analysis of recent data from Ethiopia and Pakistan to buttress his arguments. Timmer (2015) examines the same issues in a similarly applied, policy-focused manner, but adopts a more strongly macroeconomic perspective focused on structural transformation and food security, rather than the more microeconomic perspective on rural economic growth drawn upon here.

Commercially oriented smallholders are defined as farm households that produce enough from their farming to have consumption levels above the basic-needs poverty line and that, on net, sell a quarter or more of their agricultural output. The basic argument for rural economic development centered on such households is grounded in their raising their agricultural productivity through increased use of modern farming inputs and practices over an increasing share of local arable land. As the income of these local commercially oriented farm households grows, their consumption of locally produced goods and services will increase. These goods and services are those that are labor-intensive, require limited capital in their production, and typically are not marketed outside of the local community—they include construction, building repair, and associated services; transport and associated services; education, health, and other social services; furniture and handicraft-making; and food and beverage processing. A large body of research globally shows that under smallholder farming conditions, farming households will spend about half of their incremental income on such local products and services (see review in Haggblade, Hazell, and Dorosh 2007). Crucially, this mechanism for economic growth in rural communities is based on consumption linkages, not production. Local commercially oriented smallholders using modern farm inputs to increase their productivity will increase their incomes, which they will then spend in ways that propel the local economy. Such important contributions to rural economic growth are not obtained to the same extent from absentee urban-based farm owners, even if they are strongly commercially oriented (Box 5.1).

Those other rural households locally producing the goods and services that are increasingly demanded by commercial smallholders also pursue agricultural livelihoods. Given the communal land tenure system dominant in Malawi, most rural households have access to land and most use that land to grow some of their own food. However, due to capital, labor, or management constraints, or insufficient land, many are not able to generate agricultural surpluses from that land. They are poor, and depend on off-farm economic activities as much as on their farming to meet their basic needs, including for access to food.

BOX 5.1 Urban-based commercial farmers have a limited role in rural economic development

A critical aspect of this concept of rural economic development in a smallholder farming context is that it puts locally resident commercial smallholders at the center of rural economic growth. Absentee urban-based farm owners, whether operating as smallholders or at a larger estate scale, may have strong relationships with the local community and be strongly commercially oriented in their farming. Moreover, that they are farming in the area may improve all local farmers' knowledge of improved agronomic practices and new technologies, access to commercial inputs, linkages to markets, and engagement with better-capitalized traders (Deininger and Xia 2016; Burke, Jayne, and Sitko 2020). However, if these nonresident farmers are primarily consuming from urban shops and suppliers of services or imported goods and services, using the returns from their farming to finance urban consumption, their increased farm production offers only limited benefit to the local rural community. For this reason, the model of economic development used here recognizes that the local residence of commercially oriented smallholders gives them a critical role in extending economic growth across all households in a rural community. It is their local consumption that increases the incomes of producers of local goods and services.

In the household typology here, absentee urban-based farm owners fall into the category of "urban households." In any programming to support agricultural commercialization for rural economic development, program planners should be aware of the significantly lower level of local spillover effects associated with enhanced commercial production by urban-based farmers. Rural-based commercially oriented smallholder households should be preferentially targeted in such programs.

However, as the rural economic growth mechanism described here proceeds, experience from elsewhere shows that the returns that such households obtain from those nonfarm activities will increasingly surpass those that they can obtain from their low-productivity farming, propelling some specialization in local rural employment patterns (Mellor and Malik 2017). With sustained returns and continued investment—including judicious public investment in market support services, roads, transportation and communication services, electricity, and the like—many of these households producing goods and services for the local market will expand their activities to serve wider markets, further increasing their income and, of interest here, better ensuring their access to food. As this process of rural economic growth advances, many of these households will transition from being poor, subsistence-oriented households that engage in some farming to become nonfarming households specialized in livelihoods outside of agricultural production. In making this transition, many will no longer be poor.

The analysis later in this chapter shows that such a process of rural economic growth and transformation in employment patterns in rural communities in Malawi is now only in its initial stages, at best. The government has made and continues to make many of the investments and institutional reforms needed to propel this approach to rural economic development. But continued government engagement will be needed for many more years before significant economic transformation in rural communities is achieved.

This rural economic growth mechanism is rooted in crop productivity increases that, when extended over more arable land, will result in greater local economic output. Developing mechanisms that will allow productive commercial smallholder farming households to make use of the land of the less productive households in the community will be important to allow for expansion of more productive agriculture and sustain local economic growth. However, customary access to land should not be undermined. Those who have access to land through customary tenure need to be assured that they will have continuing access to that land by virtue of being members of their community. This land is important both as a household asset and as a component of the long-term economic safety net for household members, particularly later in life. A conversion of rural land to titled freehold tenure is not required for this rural economic growth mechanism to be effective. What is required in terms of land tenure reform is greater attention to how landholders could rent their agricultural land on an annual basis or for a period of a few years to other households in the community that can make more economically effective use of the land than the landholder. However, such rental arrangements will need to be designed in a way that makes it clear to all concerned how the landholder can regain use of the land.

Although leaseholds of agricultural land for 25 to 99 years have commonly been used in Malawi to allow for capital-intensive farming activities on a commercial scale, these are inappropriate in this case. The objective here is to enable renters to make use of modern inputs—fertilizer, seed, herbicides, and so on—on seasonal crops to increase their farm production and the productivity of the land beyond what the landholder would be able to achieve. One drawback is that such rental agreements would be too short to create much incentive for renters to make large investments in the land—such as irrigation, tree crops, terraces, and the like. But the principal strategic aim of these leaseholds would be to increase local annual field crop production for both incomes and food security.

As the long debates over land reform in Malawi attest and rising demographic pressures would suggest, there is increasing competition and social conflict around land. Although customary tenure systems remain at the core of smallholder access to land, it is unlikely that those systems in their current form can be easily safeguarded to ensure land access for all community members (Peters 2004). A recent study of agricultural land rental arrangements in Malawi found that the landlord—the household or individual renting out the land to the tenant—was consistently poorer on most measures than the tenant (Ricker-Gilbert et al. 2019). Moreover, the study found that many landlords rented out their land as a means to cope with economic stress. That landlords in such situations may end up losing their land is of concern. The proposals above for making it easier to rent out cropland to farmers who can make more productive use of it will need to address the significant potential for conflict as local rights to land are renegotiated. Local land governance mechanisms will need to be strengthened so that they can provide sufficient oversight to ensure fair arbitration for all parties in a land conflict.

In addition to land constraints to expanding production, commercial smallholders may also face constraints in mobilizing sufficient local labor, depending on the local context. Although rural communities in Malawi are characterized by underemployment of available labor in aggregate annually, rainfed farming systems demand close timing of farming operations, which puts heavy demands on labor at specific points in the cropping season. This creates labor bottlenecks that are exacerbated for commercially oriented smallholder farming households operating at scales that require labor beyond what is available in the household—at the same time as commercial farmers urgently require additional labor, their neighbors who might supply that labor have a strong economic preference to work on their own crops.

There are repeated examples in the colonial and postcolonial agricultural history of Malawi of the important constraint that an inability to mobilize sufficient labor poses for commercial farming ventures (for example, Green 2012b; Kydd and Christiansen 1982; Mapemba 1997). These include specific efforts in the late colonial period to support "master farmers" in rural communities (Kalinga 1993; Green 2012a), similar to what is being advocated here in focusing agricultural development efforts on the most productive and commercially oriented farmers.

In some contexts, such as in peri-urban areas where general wage rates are higher, it may not be possible for commercial smallholders to profitably employ wage labor because these households cannot offer competitive wages within the local economy. However, with strengthened agricultural markets and effective use of modern agricultural inputs, particularly labor-saving technologies, on crops that are comparatively well suited for a specific locale and the markets that serve it, the productivity benefits of judiciously using hired labor when required should exceed the costs of that labor in most growing seasons.

However, to be clear, the rural economic growth process described here is not dependent upon increased use of local wage labor in agriculture. Rather, it is driven by the increased local consumption made possible by increased crop productivity stemming from the use of modern inputs. Increased incomes will be realized by those farmers with greater production who sell their crops into reliable markets. Depending on the characteristics of the crops being produced and the local farming system, this increased production may not necessarily require any more labor than the commercially oriented smallholder farming household can supply on its own. Although some growth can be expected in the share of local workers who find primarily seasonal wage labor on neighboring farms, the more significant change in local labor markets will be a sustained increase in the share of workers engaged in remunerative activities outside of direct agricultural production.

What has been sketched here is a community-focused, long-term rural economic growth strategy based on the use of agricultural technologies and effective crop management to raise agricultural production, and strengthened markets to provide reliable incentives for commercial agricultural production. Consequently, the government does not have specific and targeted supportive roles to play vis-à-vis commercially oriented smallholder farming households—that is, specific projects to target commercial smallholder farming households are not necessarily needed. More important is the continued supply of those public goods that enable these smallholders to succeed: significantly improved and expanded agricultural research and information provision, improved transport and communication services, electricity, business skills formation, and a supportive environment for the commercial success of cooperatives and other farmers' organizations, among others. Equally important is that the government make parallel investments to strengthen agricultural markets, including through expanding trade opportunities both domestically and regionally. None of these public investment priorities are new.

Continue to safeguard the food security of other rural households while enabling them to derive sustainable livelihoods outside of agriculture

Although the central longer-term recommendation on strengthening agriculture in Malawi for improved food security is to focus on commercially oriented smallholder farming households, the government of Malawi should continue to support the economic development of other rural households. These households, which typically engage in some subsistence farming alongside other economic pursuits, make up the largest share of rural households across Malawi. With any drought, flood, crop disease, or pest infestation, their food security is at considerably more risk than is that of commercially oriented smallholders. The government has a continuing duty to ensure that the food needs of these households are met. However, the longer-term economic development perspective that the government adopts for such households should go well beyond farming. Government agencies from several sectors should work with these households to build their abilities and expertise and to put in place an economic environment in which they can sustainably obtain sufficient and expanding livelihoods from nonfarm activities.

Continued public investment is required to ensure that any farming these households engage in is as productive as possible. However, at the same time, government will need to act strategically so that such households, over the coming decades, increasingly find that their food security and overall livelihoods are best ensured through engaging in more specialized work outside of farming. This will require that government more clearly target its programs in the agriculture sector. Those programs that are focused on food security and improving the quality of the diets of rural households should be universal. All people who farm to produce some of the food they consume should benefit from such programs. However, where public investments are made to expand the participation of smallholders in agricultural markets, more targeted approaches focused on the commercially oriented smallholders make sense.

Meeting the food needs of households that are not economically productive

However, not all Malawian households can meet their food and nutritional needs through their own agricultural production, income generation through commercial agriculture, or increased engagement in nonfarm employment. Those households that are not economically productive are highly vulnerable to food insecurity. Strategically associating food security for households in

Malawi solely with agriculture or with other employment neglects these non-productive households. The food security of the most vulnerable in Malawi can be best addressed through specific policies and action on social protection (World Bank 2018). Two particular groups face permanent or temporary constraints, respectively, on their economic productivity and will require some type of support through social protection programs if they are to have reliable access to food:

1. The chronically poor who, due to old age, orphanhood, or physical or mental infirmity, are unable to be economically productive and who may not be able to rely upon others in their family or community to meet their food and other nutritional needs

2. Those who have experienced a significant adverse shock to their livelihoods, resulting in household food insecurity and an inability to provide for all of the nutritional needs of household members for a period of time as they work to rebuild their livelihoods

For the first group, given that such households are not economically active, cash transfers will likely be the best form of support. The amount each beneficiary receives should bridge the gap between their current consumption level and what would be required to purchase the food and other nutritional goods and services they need to be food secure and well nourished. Although traditional social safety nets provided by extended family or local community institutions continue to play important, even dominant, roles in meeting the needs of these chronically poor households, the government has increasingly recognized that it must provide leadership on social protection. The proper balance in support between that provided by government and that from these other traditional sources of social support will be determined, in part, by the resources that government can make available for these households.

The government has not invested significantly in such programs, however. Between 2011 and 2016, not considering the FISP and humanitarian responses to widespread acute food insecurity, funds equivalent to 0.6 percent of Malawi's GDP were allocated to safety net programs—much less than the average of 1.2 percent among countries across Africa. More than 90 percent of this funding came from Malawi's development partners (World Bank 2018). Moreover, the safety net programs are not well targeted—the poorest 20 percent of households in Malawi received only 18 percent of total spending on social safety nets (Beegle, Coudouel, and Monsalve 2018). Comprehensive food security in Malawi will not be achieved without more widespread and sufficiently funded programs of support, targeted toward those who are not economically productive.

Households that are recovering from a significant temporary adverse shock to their livelihoods are especially vulnerable to food insecurity, with the risk that some household members may become acutely malnourished. Support to enable them to meet their food and nutritional needs over the short term is required. This help may include distribution of food, cash, or vouchers to enable households to meet their immediate needs—support that the government, with close engagement by its development partners, has demonstrated it is able to provide (Babu et al. 2018). However, in addition, continuing assistance over the medium term will enable these households to regain their preshock welfare levels and, ideally, to be better prepared and more resilient in the face of any future shocks to their livelihoods.

There is a relatively good understanding on the part of the Malawi government and its partners regarding the design, targeting, and implementation of programs to help households adversely affected by broad shocks, such as droughts, that affect many people in an area. More challenging, however, is how the government and its partners can best identify and assist those households affected by more idiosyncratic shocks affecting just a single household or a handful of them. Shocks due to chronic illness or death or to fire, theft, or other destruction of household assets affecting a single household are no less damaging to a household's well-being than broader shocks. Such idiosyncratic shocks have the potential to diminish a household's standard of living and economic prospects permanently. More strategic thinking by the government can help determine how best to identify such households and what institutional capacity and services, whether kinship- or community-based, public or private, including commercial insurance options, will be needed to support them so that they can recover their livelihoods and consistently meet their food needs in a more resilient fashion (World Bank 2013).

Building a Rural Development Strategy for Malawi around Commercially Oriented Smallholder Farmers: An Empirical Appraisal

The conceptual discussion above is drawn primarily from the rural economic development experiences of other emerging economies, particularly in South and Southeast Asia, where agricultural markets and the commercial orientation of rural households are considerably stronger than in many parts of Malawi. Given the continuing dominance of a subsistence orientation in the agricultural activities of rural Malawian households, how realistic is such a development approach to Malawi? Using the different categories of households described in the research literature on rural economic transformation and listed on page 113, this section presents a profile of households that fall into each category based on analysis of the nationally representative 2016/17 Malawi IHS4. The aim is to better understand how practical and, if carried out, what the scope and scale of the effort would be to implement a rural economic development strategy in Malawi focused on enhancing the agricultural production of commercially oriented farming households and expanding remunerative nonfarm livelihood opportunities for other productive households. Such a strategy would also need to ensure that nonproductive households receive social support to enable them to be food secure.

The categorization of households used here is somewhat imperfect because of both the design of the IHS4 dataset and the significantly lower rate of market engagement of farm households in Malawi than elsewhere. The data in the IHS4 cannot be used to exactly replicate the defining characteristics of the commercially oriented smallholder farmer category. The three principal characteristics of households in this category are that they reside in rural areas, they are nonpoor and so are able to meet their basic welfare needs, and they produce considerably more crop output than they consume within their own households. However, the information on crop and, specifically, maize sales and consumption is not sufficiently harmonized in the survey to determine annually whether a household is a net maize seller or a net maize purchaser. Information in the IHS on crop sales is based on the two annual cropping seasons—the main rainfed season and, for irrigated crops, the dry season. In contrast, information on food crop purchases is based on food consumption recall over the seven days prior to the interview of the household. Consequently, no reliable estimate of the annual net maize sales position of a household can be computed from the survey data. For the assessment here, therefore, households were categorized as commercially oriented if they reported selling more than a quarter of their harvested maize annually. In doing so, no consideration was made of what share of the maize consumed by a household over the previous seven days was purchased.

However, this defining characteristic for the commercially oriented smallholder farmer category is problematic because very few farming households sell a significant share of the maize that they produce. Only 6 percent of households sell more than one-quarter of their maize harvest, according to analysis of the IHS4. As discussed in the previous chapter, there are limited incentives for farmers to produce maize for sale because, due to weak markets and unpredictable government interventions in those markets, there is significant risk that returns to production of maize for sale in the market will be insufficient.

The other criterion used to identify rural households in the commercially oriented smallholder category is their poverty status—they are not poor. The poverty analysis of the IHS4 uses a basic-needs (food and nonfood) poverty line, along with the value of per capita consumption and expenditure for a household, to assess whether the household is poor (Malawi, NSO and World Bank 2018). Given the few households making significant sales of maize, we use the lower, solely food-based, ultra poverty line, rather than the higher, basic-needs poverty line, as the poverty criterion to define which households are not poor. Doing so increases the share of households in the commercially oriented smallholder category to 5.5 percent of all households rather than the 4.1 percent that would result if the higher, basic-needs poverty line were used.

Households that fall into the not economically productive category were identified based on their poverty status: they are ultra poor, with a level of consumption per capita lower than the food poverty line, and more than half of household members are nonworkers (those younger than 15 years and older than 64)—that is, more members of the household are nonworkers than are workers. For such households, systemic labor constraints, not economic shocks or other factors, likely were a significant factor in their low welfare level at the time of the survey.

The other two categories are residual categories that are distinguished from each other based on whether the household lives in a rural or an urban area. Although many urban households engage in agricultural activities primarily to help meet their own food needs, their role in accelerating rural economic growth processes is limited, as was discussed in Box 5.1.

Table 5.1 shows the results of this categorization of the IHS4 survey households, weighted to reflect all households in Malawi that fall into each. Figure 5.1 presents maps of the share of the households in each of the 4 categories in the 32 survey strata of the IHS4, comprising the 28 districts of Malawi plus the 4 major urban centers. The size of the commercially oriented smallholder category is surprisingly small and may call into question how realistic a rural economic development program centered on such households would be. Our earlier rough calculation that potentially 40 percent of rural households in Malawi could be commercially oriented smallholder households is shown to be far from the mark under current levels of agricultural production and market performance. This divergence between expectations and reality is considered more closely later in this chapter.

In terms of the location of the households in the four categories, some elements of the definitional criteria for the particular household typology used are apparent. Commercially oriented smallholders are most commonly found in the mid-altitude plateau areas of the Central and Northern regions that

TABLE 5.1 Location and poverty characteristics of households in the different economic categories, 2016/17

Characteristic	All households	Commercially oriented smallholder farmers	Other productive rural households	Not economically productive	Urban households
Population ('000s):	16,308	882	10,395	2,015	3,015
Individuals					
Households	3,797	208	2,528	351	711
Share of households in the population, %[a]	100.0	5.5	66.6	9.2	18.7
Rural Northern region	*6.8*	*6.0*	*83.8*	*10.1*	*0.0*
Rural Central region	*36.2*	*10.7*	*80.5*	*8.8*	*0.0*
Rural Southern region	*38.0*	*3.1*	*83.6*	*13.3*	*0.0*
Urban	*19.0*	*0.0*	*0.0*	*1.7*	*98.3*
Poor (value of household consumption per capita less than basic-needs poverty line), %	44.7	25.5 ***	47.6	100.0	12.8
Ultra poor (value of consumption less than food poverty line), %	15.9	0.0 ***	9.6	100.0	1.5
Housing quality—floor, roof, or walls made with modern materials, % of households	62.5	65.4 ***	56.7	45.9	90.8
Observations	*12,447*	*636*	*8,412*	*1,161*	*2,238*

Source: Author's weighted analysis of 2016/17 Malawi Integrated Household Survey 4 (Malawi, NSO 2017).

Notes: Asterisks on the estimates for "commercially oriented smallholder farmers" indicate the statistical significance of the results of a Wald test on differences in estimates between households in this category and those in the "other productive rural households" category: * = $p < 0.10$, ** = $p < 0.05$, *** = $p < 0.01$. The urban population here is defined as residents in the four major urban centers of Malawi, district headquarters towns (bomas), and other market centers with urban characteristics as determined by the National Statistical Office of Malawi.

[a] For the second panel, the statistics in the second column are column totals, and the statistics in the third to sixth columns are row totals.

FIGURE 5.1 Maps by district and major urban centers of the share of households that fall into the four economic categories

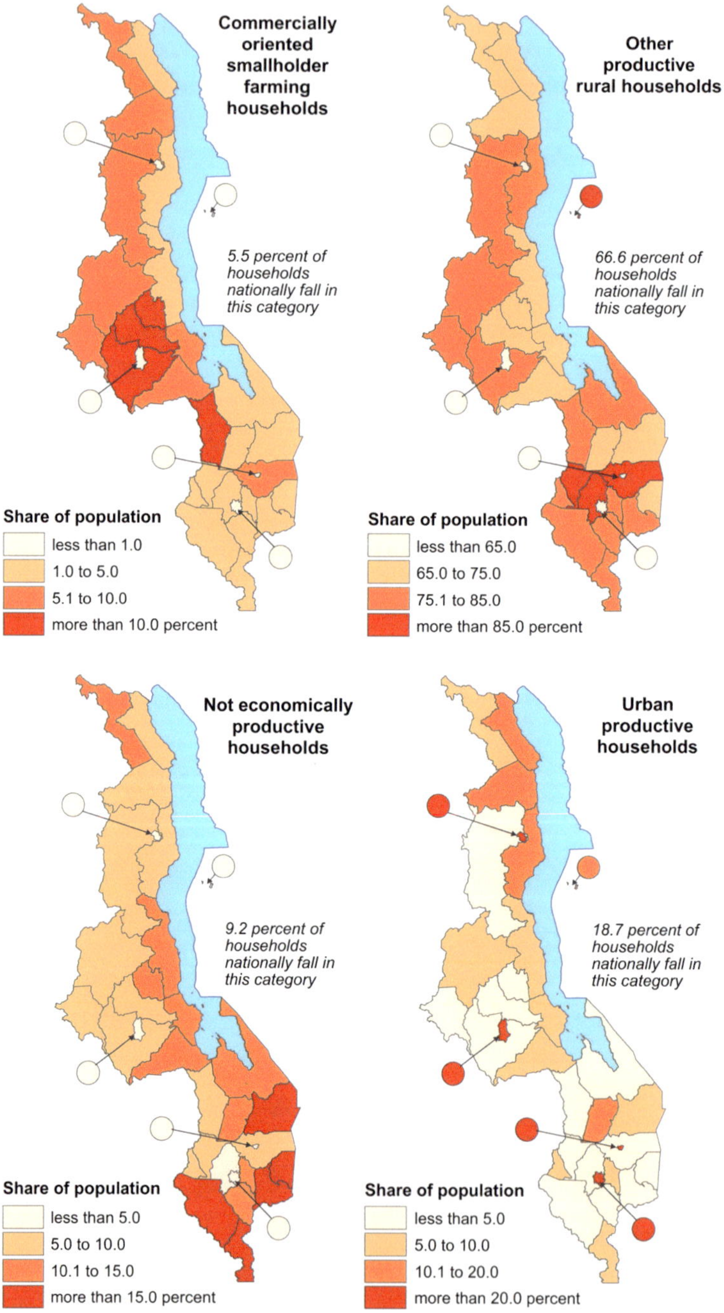

Source: Author's weighted analysis of 2016/17 Malawi Integrated Household Survey 4 (Malawi, NSO 2017).

agroecologically are among the best areas of the country to produce maize, as well as tobacco. The production of surpluses beyond household requirements is more likely in these areas than in many other parts of Malawi. That rural land pressures are also not as severe in these areas compared with upland areas in the Southern region also contributes to greater surpluses. In contrast, rural households in southern Malawi that are commercially oriented in their farm production are rare.

A larger share of households that are not economically productive is seen in the population of southern Malawi than might have been expected. This in part is an artifact of the significant drought in southern Malawi and consequent food insecurity during the IHS4 survey year of 2016/17. One of the defining criteria for such households is very low consumption levels (ultra poverty). Consumption levels for many households in southern Malawi in the survey year were sharply reduced.

Finally, a greater share of households in rural districts is made up of so-called urban productive households than might have been expected. Although such households are concentrated in the four major urban centers of Malawi, because the definition of urban used for the typology is that of the National Statistical Office, those households resident in district headquarters and rural market centers with urban characteristics fall into this category as well.

A profile of households in the four categories is presented in the several tables that follow.[1] Because the rural economic development model of interest is one centered on commercially oriented smallholder farmers, the characteristics of households in this category are highlighted. However, given the low prevalence of such households in the population, the characteristics of households in the "other productive rural households" category are also of interest. It is households in this category that are most likely to transition to a more commercial orientation in their farming, and they are also the households most likely to increasingly specialize in nonfarm livelihoods. The tables present tests of differences in estimates of the characteristics of households

1 See the appendix for additional tables that examine the households in each category in terms of their use of credit, their participation in various government social support programs, and the type of the most significant shock to their livelihoods that each experienced in the past year.

in the two categories.[2] The results of these tests provide some guidance on the opportunities, barriers, and type of support required for other productive rural households to pursue economic activities outside of farming.

The tables also present characteristics of any households that are not economically productive and of urban-based households that are economically productive. Although these households are peripheral to the rural economic transformation focus of this chapter, their characteristics are salient to considerations of food security, agricultural development, and structural transformation of Malawi's economy overall. The food needed by most households that are not economically productive will be largely provided by their extended family or other households in their community. However, the needs of many will not be fully met without support from the government. Because social safety net programs need to be part of policies and strategies for a food-secure Malawi, a better understanding of the characteristics of households that will be dependent upon them is needed.

Economically productive urban households are also of interest because many continue to engage in farming, whether in their rural areas of origin or on peri-urban plots. Households in smaller urban centers within rural areas particularly will pursue strongly agricultural livelihoods. More important, if structural transformation of Malawi's economy is to gain speed, one of the principal demographic transitions that can be expected is a movement of households from the "other productive rural households" category to the "urban households" category. If expansion in the services and industrial sectors of Malawi's economy occurs in coming years, most of the expansion in the industrial sector and much of the expansion in the service sector will be urban-based.

---

2 The results of this test are presented in asterisk form in each table. The results of statistical comparisons for the characteristics of households in the other two categories are not presented for two reasons:

- It would be challenging in these tables of descriptive statistics to comprehensively and clearly present the matrix of estimates comparison results across all four household categories.
- The principal interest of this analysis is determining empirically how feasible it would be to propel rural economic transformation on the increased agricultural productivity of commercially oriented smallholder farmers and the intensified demand of those households for the goods and services produced by households in the "other productive rural households" category.

In light of these considerations, the characteristics of households in the "not economically productive" and "urban households" categories are somewhat secondary for the analysis here. Although these characteristics are presented in the tables for completeness, the results of comparisons of estimates involving households in these categories are not presented.

Demography and education

Basic demographic characteristics of households in each category are presented in Table 5.2. Commercially oriented households are more likely to be headed by men and by younger individuals than are households in the category of "other productive rural households." These differences likely stem from differential access to resources, social and financial capital, educational attainment, and levels of entrepreneurship within the life cycle. Nonetheless, the age and sex of the household head might serve as proxy indicators, alongside several others, for targeting efforts to enhance the commercialization of household farm production, even as efforts to improve food security among rural households are more universally targeted. Households in the "not economically productive" category stand out from other households for being large with a high

TABLE 5.2 Demographic characteristics of households in the different economic categories, 2016/17

Characteristic	All households	Commercially oriented smallholder farmers	Other productive rural households	Not economically productive	Urban households
Household size, members	4.3	4.2	4.1	5.7	4.2
Dependents (younger than 15 or older than 64 years of age)	2.1	2.0	2.0	3.7	1.8
Dependents to household size ratio, mean	*0.46*	*0.44*	*0.46*	*0.68*	*0.37*
Household head age, mean years	43.1	41.5 ***	43.9	43.0	40.9
Under 35 years of age, %	*36.0*	*39.8 **	*35.7*	*33.4*	*37.2*
35 to 64 years of age, %	*50.8*	*50.7*	*49.1*	*52.9*	*55.6*
Over 64 years of age, %	*13.2*	*9.5 ****	*15.1*	*13.7*	*7.2*
Female-headed households, %	29.1	21.0 ***	30.1	41.5	21.9
Observations	*12,447*	*636*	*8,412*	*1,161*	*2,238*

Source: Author's weighted analysis of 2016/17 Malawi Integrated Household Survey 4 (Malawi, NSO 2017).

Notes: Asterisks on the estimates for "commercially oriented smallholder farmers" indicate the statistical significance of the results of a Wald test on differences in estimates between households in this category and those in the "other productive rural households" category. * = $p < 0.10$, ** = $p < 0.05$, *** = $p < 0.01$.

share of nonworking members, which is a defining characteristic of households in this category, and for many being headed by a woman.

Educational attainment levels are higher in commercially oriented farming households than in other rural households. This pattern is seen both when considering only the household head's level of education and when considering the maximum level achieved by a household member. Moreover, Table 5.3 shows increases in the education received by the younger generation—maximum education levels within households are markedly higher than the levels achieved by the heads of those households. This reflects in part the impact of the government's universal primary education program that has been in place since 1994. Although problems related to the quality of instruction and facilities remain, improved access to schooling has increased average educational attainment. The average years of education successfully completed for those ages 15 to 24 years increased from 5.0 to 6.2 years between the 1998 and 2008

TABLE 5.3 Educational attainment within households in the different economic categories, 2016/17

Characteristic	All households	Commercially oriented smallholder farmers	Other productive rural households	Not economically productive	Urban households
Educational attainment of household head, % of households					
No formal education	17.2	8.2***	20.1	25.3	5.6
Some primary	57.8	66.3**	61.3	67.1	38.0
Some secondary	*22.2*	*23.7****	*17.3*	*7.6*	*46.2*
Beyond secondary	2.8	1.8	1.2	0.0	10.2
Maximum educational attainment in household, % of households					
No formal education	*3.9*	*1.8****	*5.0*	*3.0*	*1.2*
Some primary	*57.5*	*57.1***	*62.8*	*83.6*	*26.3*
Some secondary	*34.9*	*37.9****	*30.7*	*13.3*	*59.5*
Beyond secondary	3.7	3.3**	1.6	0.1	12.9
Observations	*12,447*	*636*	*8,412*	*1,161*	*2,238*

Source: Author's weighted analysis of 2016/17 Malawi Integrated Household Survey 4 (Malawi, NSO 2017).

Notes: Asterisks on the estimates for "commercially oriented smallholder farmers" indicate the statistical significance of the results of a Wald test on differences in estimates between households in this category and those in the "other productive rural households" category. * = $p < 0.10$, ** = $p < 0.05$, *** = $p < 0.01$.

censuses (analysis by author). Rising levels of human capital in rural communities should prove an important driver of rural economic transformation, even if the impacts may not be seen for decades and a range of complementary investments will also be needed to advance local rural economies. Members of commercially oriented farming households are at the forefront of these improvements in education levels across Malawi.

Agricultural production and crop sales

Almost all farmers in Malawi engage in rainfed cropping (Table 5.4). However, commercially oriented smallholder households are more likely than other households to also engage in some irrigated farming. Although the share of those that have irrigable land is less than 15 percent, access to such land is a distinguishing feature of commercially oriented smallholders. Such households also have the largest holdings of cropland, with access to almost twice as much land as other rural households. Although renting-in of land is not common among farming households in Malawi, commercially oriented smallholders are more likely to do so than are other households.

Hiring-in of agriculture labor to supplement household labor is done by only one out of five households that engage in farming. However, commercially oriented smallholders, as well as urban households that farm, are more likely than other households to employ outside labor. But outside labor is not used any more intensively by commercial smallholders than by other rural households that make use of such labor.

Household members' temporarily hiring out their labor (ganyu) for any task, whether agricultural or nonagricultural, is much less common among working members of commercial smallholder households than it is among members of other rural households. Members of chronically poor, not economically productive households are most likely to rely on such employment and engage in it more intensively than do members of other households.

Characteristics of the cropping patterns of the different types of households are presented in Table 5.5. The characteristics of households that engage in any sales of maize closely match the criteria used to define commercially oriented smallholder farming households. Other rural households are unlikely to sell any of the maize that they produce. However, more than a quarter of urban households that engage in maize production report selling some of their harvest. Both commercially oriented smallholders and urban households that produce maize are likely to use improved seed (hybrids or open-pollinated varieties), whereas households in the other two categories that produce any maize are as likely to use local maize varieties as to use improved varieties.

TABLE 5.4 Agriculture-related characteristics of households in the different economic categories, 2016/17

Characteristic	All households	Commercially oriented smallholder farmers	Other productive rural households	Not economically productive	Urban households
Engaged in any crop production in previous year, %	76.2	100.0***	85.3	91.0	30.1
Of those that did so, rainfed only	90.9	86.6**	90.8	93.9	92.0
Of those that did so, irrigated only	0.7	0.5	0.6	0.6	*2.0*
Of those that did so, both irrigated and rainfed	8.3	12.9**	8.6	5.4	6.0
Cropland area used in past season, ha					
Average, all households	0.52	1.01***	0.57	0.50	0.17
Average, all households that produced crops	0.65	1.01***	0.65	0.52	0.53
Per household member, all households	0.14	0.28***	0.16	0.09	0.05
Per household member, all households that produced crops	0.17	0.28***	0.18	0.09	0.14
Rent in some land, % of all households	7.8	12.1**	8.3	5.8	5.6
Cropland rented in, ha, average for households renting in	0.52	0.69	0.53	0.40	0.43
Rent out some land, % of all households	1.2	1.8	1.3	1.4	*0.5*
Hire in agricultural labor, % of households that produced crops	19.1	30.2***	16.1	7.6	*47.2*
Amount of hired-in labor, for those hiring in, hours/ha cropland over past cropping year	26.7	23.9	25.5	18.2	*35.5*
Members engage in any hired-out temporary work (ganyu), either agriculture or nonagriculture, %	60.3	57.1***	65.2	81.7	33.3
Amount of labor hired out over past year for those hiring out, hours per worker in household	57.9	46.7***	55.9	68.4	*64.4*
Observations	*12,447*	*636*	*8,412*	*1,161*	*2,238*

Source: Author's weighted analysis of 2016/17 Malawi Integrated Household Survey 4 (Malawi, NSO 2017).

Note: Asterisks on the estimates for "commercially oriented smallholder farmers" indicate the statistical significance of the results of a Wald test on differences in estimates between households in this category and those in the "other productive rural households" category. * = $p < 0.10$, ** = $p < 0.05$, *** = $p < 0.01$.

TABLE 5.5 Crop production and sales of households in the different economic categories, 2016/17

Characteristic	All households	Commercially oriented smallholder farmers	Other productive rural households	Not economically productive	Urban households
Maize (any) production, % of households	72.0	100.0***	80.4	83.2	28.7
Sold any maize, % of households	11.6	100.0***	6.9	5.3	5.3
Share of maize harvest sold, for households that sold any, %	34.6	50.1***	15.5	28.3	41.1
Local maize production, % of households	34.1	34.6*	39.1	45.3	10.4
Improved maize production, % of households	39.6	68.9***	43.1	39.2	18.7
Rice production, % of households	3.3	1.5***	3.9	3.6	1.3
Other grain production, % of households	7.7	4.2***	8.9	14.1	1.3
Cassava production, % of households	4.5	4.2	5.5	5.0	0.8
Sweet potato production, % of households	2.3	3.6	2.6	2.0	0.8
Irish potato production, % of households	0.8	2.2**	0.9	0.3	0.1
Groundnut production, % of households	11.8	22.6***	13.7	8.5	3.2
Sold any groundnut, % of households	6.8	18.0***	7.7	4.5	1.3
Share of groundnut harvest sold, for households that sold any, %	58.4	63.3*	57.4	53.3	68.9
Bean and cowpea production, % of households	10.7	25.3***	11.0	9.3	5.7
Sold any bean and cowpea, % of households	3.3	10.3***	3.4	2.1	1.4
Share of bean and cowpea harvest sold, for housenolds that sold any, %	66.8	68.6	66.1	70.3	66.0
Pigeonpea production, % of households	17.0	11.1***	20.0	24.5	4.6
Sold any pigeonpea, % of households	6.0	6.6	7.1	7.3	1.3
Share of pigeonpea harvest sold, for households that sold any, %	59.0	65.6	59.2	57.4	51.1
Soyabean production, % of households	7.7	16.3***	8.6	6.6	2.7
Sunflower production, % of households	0.9	2.2**	0.9	1.3	0.1
Tobacco production, % of households	4.9	11.8***	5.6	3.8	0.7
Cotton production, % of households	1.2	0.4***	1.6	1.5	0.2
Observations	*12,447*	*636*	*8,412*	*1,161*	*2,238*

Source: Author's weighted analysis of 2016/17 Malawi Integrated Household Survey 4 (Malawi, NSO 2017).

Note: Asterisks on the estimates for "commercially oriented smallholder farmers" indicate the statistical significance of the results of a Wald test on differences in estimates between households in this category and those in the "other productive rural households" category. * = $p < 0.10$, ** = $p < 0.05$, *** = $p < 0.01$.

Few households in any category produce both local and improved varieties of maize.

Commercially oriented smallholders are less likely than other productive rural households to produce rice or other cereals. These results in part reflect the maize-centric definition of commercially oriented smallholders. Rice is more common in alluvial and lakeshore areas where maize is not as well adapted for production as it is elsewhere in the country, whereas sorghum and pearl millet are somewhat better adapted than maize for the Lower Shire Valley as well as in alluvial areas on the margins of the Shire Highlands in southern Malawi. Local consumption preferences will also be a factor in the cropping patterns for these other cereals.

The share of households that produce any roots and tubers is low. In this regard, the IHS4 data contrast sharply with the agricultural production estimates of the Ministry of Agriculture, which showed the total area planted in roots and tubers was more than one-third of the area planted in maize in 2019.[3] Little difference is apparent in the share of rural households that produce cassava and sweet potato across categories. However, Irish potato is more likely to be produced by commercially oriented smallholders than by households in the other categories, reflecting the higher commercial demand for Irish potatoes compared with sweet potatoes and cassava.

Groundnuts, beans, and cowpeas are more likely to be produced by commercially oriented smallholders than by households in other categories. However, these crops remain important for own consumption—while more than half of households that produce groundnuts will sell some of their harvest, those that do will still keep about 40 percent for home use. Less than one-third of bean and cowpea producers sell these crops, but if they do, they sell about two-thirds of what they harvest. Pigeonpea is more commonly produced by households in the "other productive rural" category. This reflects in part the geography of the production of pigeonpea, which is predominantly produced in southern Malawi, where commercial production of maize is not common (see Table 4.1). Less than half of pigeonpea-producing households sell any of their harvest, but those that do so sell about 60 percent of what they harvest. Groundnut, bean, cowpea, and pigeonpea are significant for households engaged in any agricultural production in that they are important for meeting household food needs, but also can be readily sold to meet household cash needs. In this regard, these crops are similar to maize.

---

3 On the challenges of estimating agricultural production statistics for cassava, see discussion in footnote 3 in Chapter 3.

Soyabean, sunflower, tobacco, and cotton are cash crops. Commercially oriented smallholders are significantly more likely than households in other categories to produce soyabean, sunflower, and tobacco. The IHS data show that more households produce soyabean than tobacco, possibly reflecting lower barriers to the production and marketing of soyabean compared with tobacco. Although only a few households produce cotton, those that do are less likely to be in the commercially oriented smallholder category. This fact principally reflects the geography of cotton production in Malawi, which is grown where maize is less well adapted agroecologically, so that commercial production of maize is not common in cotton-growing areas.

Commercially oriented smallholders are significantly more likely to own cattle, goats, sheep, and pigs than are households in other categories and, of those that own these animals, to own larger numbers of them (Table 5.6). Poultry ownership is common across all household categories, with urban households more likely than rural households to own chickens and other fowl.

TABLE 5.6 Livestock ownership of households in the different economic categories, 2016/17

Characteristic	All households	Commercially oriented smallholder farmers	Other productive rural households	Not economically productive	Urban households
Own livestock, % of households	33.9	55.0***	37.8	28.5	16.7
Tropical livestock units owned, of those owning	0.64	0.99**	0.65	0.30	0.51
Own cattle, % of households owning livestock	8.7	14.7**	9.0	4.7	4.1
Own goats, sheep, or pigs, % of households owning livestock	54.1	66.2***	56.0	50.3	30.1
Own poultry, % of households owning livestock	64.9	59.8	64.2	59.6	79.4
Observations	*12,447*	*636*	*8,412*	*1,161*	*2,238*

Source: Author's weighted analysis of 2016/17 Malawi Integrated Household Survey 4 (Malawi, NSO 2017).

Note: Asterisks on the estimates for "commercially oriented smallholder farmers" indicate the statistical significance of the results of a Wald test on differences in estimates between households in this category and those in the "other productive rural households" category. * = $p < 0.10$, ** = $p < 0.05$, *** = $p < 0.01$. Tropical livestock units (TLUs) are based on the following conversion factors: poultry = 0.01 TLU; calf = 0.3; steer or heifer = 0.7; cow = 0.7; ox or bull = 0.8; donkey, mule, or horse = 0.6; goat = 0.1; sheep = 0.1; pig = 0.2.

However, cattle ownership is quite rare across all households, with less than 10 percent of households having any. The limited landholding size for most rural households imposes important constraints on cattle ownership, whereas goats, sheep, and pigs can be raised more effectively on the small areas.

Nonagricultural income sources

Few differences are seen in participation in nonagricultural economic pursuits between categories of rural productive households (Table 5.7). Members of commercially oriented smallholder households and other productive rural households both have low levels of participation in wage employment. Wage employment, as one would expect, is most common for members of urban

TABLE 5.7 Nonagricultural income sources and credit use of households in the different economic categories, 2016/17

Characteristic	All households	Commercially oriented smallholder farmers	Other productive rural households	Not economically productive	Urban households
Have any members with wage employment, % of households	18.8	11.3	11.9	5.3	52.2
Wage employment under long-term contract or permanent, % of households with members with wage employment	*70.9*	*69.3*	*63.2*	*44.2*	*78.6*
Have members engaged in household enterprise, %	26.6	27.5*	23.6	14.1	43.0
Of which at least one of household's enterprises are permanently operating (not seasonal), %	*28.9*	*28.6*	*24.2*	*22.0*	*39.2*
Of which at least one of household's enterprises requires skills to produce merchandise or services offered (not petty production or trading), %	*15.0*	*12.6*	*13.8*	*15.3*	*17.7*
Of which at least one of household's enterprises has employed labor from outside the household in past year, %	*9.3*	*10.9*	*6.8*	*4.1*	*14.7*
Have member who receives regular income payment, for example, pension, remittances, real estate or asset rentals or sales, %	43.3	44.4	40.9	32.8	56.8
Observations	*12,447*	*636*	*8,412*	*1,161*	*2,238*

Source: Author's weighted analysis of 2016/17 Malawi Integrated Household Survey 4 (Malawi, NSO 2017).

Notes: Asterisks on the estimates for "commercially oriented smallholder farmers" indicate the statistical significance of the results of a Wald test on differences in estimates between households in this category and those in the "other productive rural households" category. * = $p < 0.10$, ** = $p < 0.05$, *** = $p < 0.01$.

households. Commercially oriented households are more likely than other productive rural households to have members engaged in household enterprises. However, only about one-third of households with enterprises operate them permanently—most are seasonal activities. This suggests that few rural households are specializing in such enterprises as part of a transition out of agricultural production. Moreover, the household enterprises are not qualitatively different across the four categories of households. Most such enterprises, regardless of what category of household operates them, primarily involve petty production or trading and do not require skilled labor.

Of importance to the strategy for rural economic development advocated in this chapter, it appears that other productive rural households are not more likely than commercially oriented smallholder farming households to be working in nonfarm enterprises producing goods and services that are primarily for local consumption. This includes employment in construction, building repair, and associated services; transport and associated services; education, health, and other social services; furniture and handicraft-making; food and beverage processing; and the like. One of the dynamic elements of the strategy is that increasingly these households will seek their livelihoods in these activities, relying on the commercially oriented farming households to supply the food they require through the market, and reducing their dependence on subsistence farming to meet their food needs. However, the fact that we are not seeing a larger number of other productive rural households working outside of agriculture suggests that the desired rural economic transformation process has not yet started.

Food consumption

Finally, food consumption patterns of households in each category are examined in Table 5.8. Although commercially oriented rural households are more likely than others to have consumed maize in the past week, such levels are high for all households, so any differences between categories in what share of households consume maize are not practically significant.

However, there are important differences in the sources of maize consumed across household categories. Commercially oriented households are much less likely than other households to have bought the maize they consume and much more likely to have produced that maize themselves. This pattern highlights the continued weakness of Malawi food markets and broader agricultural markets, and the high risk that consumers continue to perceive in relying on those markets for their staple food. It also suggests that although

TABLE 5.8 Source of maize consumed, dietary diversity, and experience of recent food insecurity of households in the different economic categories, 2016/17

Characteristic	All households	Commercially oriented smallholder farmers	Other productive rural households	Not economically productive	Urban households
Maize consumption in past week, % of households	97.8	99.4***	97.8	95.5	98.7
Maize consumed per capita in past week, kg	2.8	3.1**	3.0	1.8	2.6
Purchased, share of maize consumed for those consuming, %	53.5	30.3***	47.4	58.3	79.7
Own produced, share of maize consumed, %	34.1	61.5***	38.5	26.0	14.4
Gift, share of maize consumed, %	12.3	8.1***	14.1	15.8	5.9
Household Dietary Diversity Score (consumption in past 7 days out of 12 food groups), mean	7.7	8.3***	7.4	5.5	9.9
Experienced food insecurity within household:					
In past 7 days, %	63.4	49.0***	66.1	85.2	47.2
In past 12 months, %	72.5	60.9***	78.5	91.8	45.1
Observations	*12,447*	*636*	*8,412*	*1,161*	*2,238*

Source: Author's weighted analysis of 2016/17 Malawi Integrated Household Survey 4 (Malawi, NSO 2017).

Notes: Asterisks on the estimates for "commercially oriented smallholder farmers" indicate the statistical significance of the results of a Wald test on differences in estimates between households in this category and those in the "other productive rural households" category. * = $p < 0.10$, ** = $p < 0.05$, *** = $p < 0.01$.

some farming households may be quite commercial in how they plan their production, this should not be taken to mean that they are specialized producers that are exploiting their comparative advantage and relying on the market to supply the food, goods, and other services for which they do not have a comparative advantage. Such households continue to place a high value on meeting their own needs directly and insulating themselves from any market-related risks to their subsistence.

Overall, we see that few rural households are not subsistence oriented and, rather, rely on the market and specialized producers of those goods and services they require. This is not to say that these households are autarkic, but no systematic division of specialized labor is apparent in rural communities. An important goal of the broader development program advocated here is increased economic interdependence among rural households, mediated by efficient markets. Based on this analysis of household production and consumption patterns and the role of the market in them, we see more evidence that rural Malawi appears to be closer to the start of the pathway to this goal than to its achievement.

The last rows of Table 5.8 examine dietary diversity and food insecurity. Commercially oriented households have more diverse diets and are significantly less likely than other rural households to have experienced food insecurity in the past week or past year. However, urban households perform better than commercially oriented smallholders on both sets of measures.

Changes in distribution of households across economic categories between 2004/05 and 2016/17

This chapter argues that agricultural and rural economic development strategies in Malawi should focus on commercially oriented smallholder farming households. However, the strength of this argument is undermined by the small share of Malawian households that fall into this category—only 5.5 percent in 2016/17. In our conceptual discussion in the chapter's opening paragraphs, we estimated that potentially up to 40 percent of rural households in Malawi may be commercially oriented smallholder households. On the face of it, it seems unlikely that significant improvements in household livelihoods and in the performance of the Malawian economy can emerge from efforts to increase the role that such a small group of rural households play in their local economies.

A better understanding of how the relative size of these household groups is changing over time can shed some light on this issue. If the share of Malawi's farming households that are commercially oriented shows a declining trend, it could be concluded that the development strategy advocated in this chapter might have worked in the past when more farming households were actively engaged with output markets, but that the development opportunity that situation presented has since been lost. In contrast, if the share of commercially oriented farming households has remained low but steady, it suggests that insufficient attention has been paid to building the capacity of these households to engage in higher-productivity commercial agricultural production, and to improving the enabling environment for their financial success, particularly through improved markets.

To better understand trends in the relative sizes of the four household economic categories, the typology used for the analysis of the IHS4 survey households was applied to households surveyed in the second IHS of 2004/05, and the third IHS of 2010/11. The population-weighted results of these analyses are presented in Table 5.9. What we find is that the share of commercially oriented farming households among all Malawian households has remained low but relatively steady over the three survey rounds. The only consistent trend across the three survey rounds is the increasing share of the population made up by economically productive households in urban areas. This is consistent with rising urbanization in Malawi. At a regional level, there are indications that rural households in the Central region are becoming more commercially oriented, whereas rural households in the Southern region are becoming less so. However, because the small share of commercially oriented rural households in the south in 2016/17 may also reflect the poor cropping conditions

TABLE 5.9 Households in the different economic categories in 2004/05, 2010/11, and 2016/17, by rural regions and urban, weighted percentage share of households

Characteristic	All households	Commercially oriented smallholder farmers	Other productive rural households	Not economically productive	Urban households
2004/05 (IHS2)	**100.0**	**6.3**	**71.8**	**10.3**	**11.7**
Rural Northern	9.4	7.5	81.7	10.8	0.0
Rural Central	36.2	8.2	83.8	7.9	0.0
Rural Southern	42.5	6.1	79.5	14.4	0.0
Urban	12.0	0.0	0.0	2.5	97.5
Observations	*11,280*	*687*	*8,029*	*1,165*	*1,399*
2010/11 (IHS3)	**100.0**	**4.4**	**68.0**	**12.3**	**15.3**
Rural Northern	10.8	5.6	80.6	13.8	0.0
Rural Central	34.1	7.1	81.7	11.2	0.0
Rural Southern	39.5	3.4	79.7	16.9	0.0
Urban	15.6	0.0	0.0	2.2	97.9
Observations	*12,271*	*556*	*8,098*	*1,437*	*2,180*
2016/17 (IHS4)	**100.0**	**5.5**	**66.6**	**9.2**	**18.7**
Rural Northern	6.8	6.0	83.8	10.1	0.0
Rural Central	36.2	10.7	80.5	8.8	0.0
Rural Southern	38.0	3.1	83.6	13.3	0.0
Urban	19.1	0.0	0.0	1.7	98.3
Observations	*12,447*	*636*	*8,412*	*1,161*	*2,238*

Source: Author's weighted analysis of 2004/05 (IHS2), 2010/11 (IHS3), and 2016/17 (IHS4) Malawi Integrated Household Surveys (Malawi, NSO and World Bank 2007a; Malawi, NSO 2012, 2017).

Note: Statistics in second column are column totals, and those in the third to sixth columns are row totals.

there during the IHS4 survey year, these regional patterns should be confirmed using future survey data.

Commercially oriented smallholder farming households, though relatively few, are not withering away over time, despite an agriculture sector that overall is facing shrinking landholdings and continuing stagnant levels of agricultural productivity. Consequently, we cannot conclude that a window of opportunity for rural economic development through focusing on commercially oriented smallholders has now closed.

Assessment

The longer-term prospects for food security and rural economic transformation require engagement in fostering the economic success and growth of the commercially oriented smallholder category of rural households, who will form strong economic linkages with their neighbors who engage in increasingly remunerative nonagricultural economic activities. That so few rural households are well integrated into markets reflects continuing weakness in Malawi's markets and might seem to threaten the viability of this rural development strategy. However, the changes such a strategy for rural economic development would require—greater market integration for all rural households and higher agricultural productivity—are essential if rural Malawi is to avoid economic stagnation and increased immiseration. Although the target group for such a strategy—commercially oriented smallholders—currently is quite small, with effective market development and increased agricultural productivity, its size as a share of rural households will grow. At the same time, other rural households will see many of the barriers and risks to expanding nonagricultural livelihoods become less constraining.

For development program planning, the criteria used here to define the household typology can be refined. In the discussion of patterns in the typology profile tables, some weaknesses in the analytical framework were recognized, primarily linked to defining commercial farming solely based on maize sales. An alternative criterion could be used. For example, AGRA (Alliance for a Green Revolution in Africa) has explored a more disaggregated rural household typology to guide the design and targeting of agricultural development efforts (Hazell 2017; Hazell et al. 2017; Biscaye, Anderson, and Reynolds 2017). For farm household modeling purposes, other researchers studying Malawi's agriculture sector have created even more complex household typologies (Dorward 2002, 2006; Douillet and Toulon 2014). These alternative household typologies can provide insights for the design of efforts to transform agriculture and the rural economy in Malawi, in addition to those gained with the simple typology here.

Nonetheless, we can expect that using these alternative typologies to understand the general agricultural and rural economic context will likewise show that agricultural markets in Malawi remain too weak to foster a strong commercial orientation among smallholder farmers, particularly in the production of food crops. The consequence of so few farmers engaging in the commercial production of maize or other food crops is a higher risk of food insecurity for many households and an increasing incidence of food crises for the nation as a whole.

Policy Strategies for Rural Economic Transformation

As plans for agricultural development in Malawi are executed, targeting should be done in a manner that best enables the attainment of the longer-term objectives of each component of these plans. Those components that are focused on food security and improving the quality of the diets of rural households should be universal. All those who farm to produce some of the food they consume—both nonpoor, commercially oriented smallholder farming households and poor, subsistence-oriented households—should benefit from such programs. These universal programs would include agricultural research and extension services, dedicated agriculture-for-nutrition and other food security programs, efforts to strengthen rights to land, activities for management of pests and diseases as well as other disasters, and others.

However, targeted approaches will be required for any public investments or programs designed to expand the participation of smallholder farming households in agricultural markets or for which commercial viability is inherent to the success of the effort. Activities that should be targeted to commercially oriented smallholders include those aimed at creating self-sustaining cooperatives and other farmer groups, bringing into production somewhat more capital-intensive irrigation schemes, or improving financial access to modern farm inputs on a full-cost, commercial basis. The greatest impact of such market-centered activities will be realized if they are principally targeted to nonpoor, commercially oriented smallholder farming households.

Making such targeting decisions will require strong commitment and leadership to surmount the political challenges that will arise from the government being selective in how its resources are deployed in the agriculture sector.

However, if the food security of Malawi is to be better ensured and sustained for future generations, agricultural transformation is needed. Reaching this goal will require the government to target its resources to catalyze such transformation specifically toward farmers who are able to multiply the benefits of such public investments. Allocating resources for these purposes to other rural households is misplaced.

At the same time, government can work to improve the prospects for these other households to pursue sustainable and remunerative nonagricultural livelihoods in the long term by building their capacity so that they can successfully and profitably operate nonfarm enterprises or find jobs that provide a good wage. To this end, the government should support specialized training to improve the ability of members of the increasingly nonagricultural rural households to offer goods and services of high quality to local consumers, to develop marketable technical skills, and to build their entrepreneurship and business management skills.

But even as efforts are made to provide good incentives to these other rural households to exit agriculture and pursue their livelihoods elsewhere, government must be vigilant in ensuring that these households are able to access the food they require. This includes continuing to provide support to poor households that farm so that they are as productive as possible. Government will need to remain vigilant in monitoring local food security situations and be ready to assist vulnerable households both through targeted support, using market mechanisms as far as possible, and if the scale of the crisis warrants, through humanitarian assistance.

This differentiated vision of rural communities will require a clear division in the priorities of the Ministry of Agriculture. The ministry will still support all rural households that engage in some farming, even if not commercially oriented, to make the most effective use of their land to ensure their own food security in the short to medium term. However, in fulfilling its mandate to support the longer-term development of Malawi's agriculture sector, the ministry should increasingly orient its efforts toward promoting the economic success of those with sufficient resources to use agricultural production as an engine of local economic growth—commercially oriented smallholder farming households. Somewhat more narrow targeting of the different government programs in the agriculture sector will be needed to better reflect the sharply different economic development and agricultural transformation potential among Malawian households that farm.

Although there are trade-offs in determining how best to allocate public resources both to achieve food security, on the one hand, and to propel rural economic change by expanding the participation of smallholder farming households in agricultural markets, on the other, there are a broad range of public investments that will support both objectives. These are the specific actions the government can take to create an enabling environment for both increased commercialization of agriculture and expanded opportunities for nonfarm employment. In doing so, the government can increase access to food for all. These actions include investing in both basic education and specialized technical training; expanding the network of all-season rural roads; extending the rural electricity supply; providing increased support to agricultural research and extension services; and continuing efforts to improve communication, particularly to support marketing and trade. Such investments, though initially they may primarily benefit more commercially oriented farm households, will provide a foundation for economic change in rural communities by expanding local nonfarm employment opportunities as well.

In summary, in terms of the economic categories of households that make up the rural population in Malawi, these efforts aim to lead toward a long-term vision for rural economic development with the following characteristics:

- Most agricultural production will be done by a significant number of productive, commercially oriented smallholder farming households, even if these productive farming households make up a much smaller share of all rural households engaged in any farming than is now the case. As such farming households become more specialized, become even more market oriented, and farm larger plots (ideally renting land from poorer neighbors), they will supply the food that increasing numbers of Malawians will obtain from the market.

- This agricultural transformation will be associated with a sharp increase in the share of the population made up of nonfarming households specialized in livelihoods outside of agricultural production, whether locally in their rural communities or elsewhere. The food needs of these households will increasingly be supplied through strengthened agricultural markets across the country.

- The food and other basic needs of the chronically poor in rural communities, and elsewhere, will be effectively met in part through social safety nets.

- The share in the rural population of poor, subsistence-oriented households that engage in some farming will shrink over time. If this vision of rural economic transformation is successful, it no longer will make sense for any household to engage in subsistence agricultural production to meet its food needs.

The public goods, services, and oversight that government provides can enable commercial smallholders to succeed economically while also fostering the economic advancement of their poorer neighbors. These other economically productive rural households will increasingly rely on the provision of labor and nonfarm goods and services, rather than subsistence farming, for their livelihoods and to ensure their own food security. Yet, regardless of the local success of such a rural economic growth strategy, social protection services still will be required for those households that are not economically productive and that may not be able to rely upon others in the community to meet their needs. However, we also must be realistic about the limited extent to which social protection support can be scaled up in communities across Malawi.

Finally, the discussion here neglects a third type of farmer—the large-scale commercial farmer. Such farmers are generally not residents in rural communities of Malawi, so in the typology used here, they are categorized under "urban households." From an agricultural transformation perspective, as discussed in Box 5.1, such households are not expected to contribute significantly to accelerating the rural economic transformation mechanism centered on increased demand for locally produced goods and services. If the government seeks broad and inclusive economic structural transformation in rural Malawi, there is little reason for it to invest much, if any, of its resources in such large-scale farmers, beyond putting in place the components of an enabling environment for their business activities, including strengthened markets and supportive commercial regulations. However, government should restrict the ability of large-scale commercial farmers to expand their operations at the expense of smallholders, such as through land expropriation.

In sum, this approach to agricultural and rural economic development is oriented toward creating a significantly more diverse and integrated national economy. If it is successful, particularly in rural communities, agriculture will continue to play a dominant role in the local economy, but not as pervasively as now is the case. At the same time, increasing numbers of rural households will obtain their food from the market because it makes economic sense to do so, rather than relying on markets only when their own production falls short. Stronger markets with more reliable supplies of food from both domestic and regional sources will make purchasing food a sound choice. In this way, improved food security for all will follow.

Chapter 6

STRUCTURAL TRANSFORMATION OF MALAWI'S ECONOMY TO END HUNGER

The discussion in this book on strengthening access to food and developing the agriculture sector fits within a broader development agenda of structural transformation of Malawi's economy. Food security will be best assured if Malawians can obtain good employment of any sort, whether within or outside of farming, confident that they can use their income to obtain food in significantly more reliable markets at reasonable prices. This chapter looks at several facets of such an economic transformation.

- Using a model of Malawi's economy, the outcomes of sector-focused investment options are compared. There are important trade-offs—a services-focused development program provides the most economic growth and employment, whereas an agriculture-focused approach will be best for improving the welfare of the poor. But of concern is that continued lagging growth for Malawi's economy under an agriculture-focused approach may not provide sufficient income and employment opportunities to enable many poor Malawian households to exit poverty.
- Human capital development will be central to transforming the economy and propelling its growth. Several investment areas are discussed, including the importance of urban development to economic change.

The role of international wage labor migration in improving the well-being of Malawian households is also discussed. The use of this livelihood option will intensify as long as a structural transformation of Malawi's economy is not achieved. Strategies should be developed now so that such migration will benefit all concerned—the migrant, his or her family in Malawi, and Malawi as a whole.

In this book, the three broad recommendations for public action to sustainably reduce food insecurity in Malawi are (1) to continue to make efforts to raise agricultural productivity; (2) to sustainably deepen and otherwise strengthen domestic and export markets for all agricultural products, but particularly food markets; and (3) to more narrowly target public investments and programs for the development of the agriculture sector and the rural economy. The last recommendation involves viewing commercially oriented smallholder farmers as the future of Malawian agriculture, even as other productive rural households are assisted in securing their livelihoods through nonfarm activities. In this chapter the focus is on what is needed to expand the economic options for members of these other rural households for whom seeking employment outside of agriculture will likely be the best career path.

An expansion of remunerative employment in nonagriculture sectors of the economy will be needed to foster considerable movement of labor and capital into industry and services. This will require a structural transformation of Malawi's economy. However, bringing about such a transformation will require increased attention to raising agricultural productivity, even as a broad range of public investments are made to improve productivity in both the industry and service sectors (Barrett et al. 2017; World Bank 2019).

Enabling many Malawian households to move away from agriculture-based livelihoods is not a new recommendation. Kettlewell, the Director of Agriculture for the colonial government, in a submission on agricultural policy to the Legislative Council in 1955, stated:

> *The intensification of farming on a scale which will enable the individual farmer to enjoy a reasonable standard of living involves the natural corollary that the balance of the population must have some alternative form of employment for their livelihood. Subsistence cultivation must disappear: men must either be farmers in the true sense or sever their direct connection with the land.... There must therefore be two complementary objectives of policy in dealing with the agrarian problem: the one to intensify farming and the other to foster every feasible form of alternative occupation in order that the inefficient subsistence cultivator may be squeezed off the land and yet have the means of a reasonable livelihood. (1955, 5)*

We disagree with Kettlewell's view that employment in the manufacturing and service sectors will provide a fallback livelihood only for inefficient farmers—a view that privileges agricultural production for Malawi's economic development. Rather, government needs to put in place adequate incentives for all Malawians to find and engage in sufficiently remunerative work in any of the three sectors of the economy. With long-term attention to development of the industry and service sectors, the welfare and food security of most Malawian workers and their households will be best served through those workers leaving the agriculture sector and obtaining employment in the other sectors.

Expanding the economic options of individual Malawians will require a transformation of the structure of Malawi's economy. Such an aim is not easily achieved, and how it can be done successfully is not well understood, despite much research on the topic (for example, McMillan, Rodrik, and Sepúlveda 2017). Moreover, historical analyses have shown that it is not easy to define the specific role government should play in economic transformation efforts. Here only broad pointers are outlined for how the transformation of Malawi's economy might be accelerated.

Sectoral Investment Patterns and Structural Change in Malawi's Economy

A necessary driver in the transformation of any economy is increased economic productivity, including in agriculture. Increased output will generate surpluses that can be used to finance additional production, foster new enterprises, and provide jobs offering higher returns to more workers. A dynamic computable general equilibrium (CGE) model of the economy of Malawi has been used to better understand the development gains that might be realized by 2030 through significantly increasing the productivity of each of the three sectors of the economy—agriculture, industry, and services (Benson and Hartley 2020).[1]

CGE models are useful for gaining insight into the likely impact of sectoral changes in the economy. They are designed to capture the functioning of a market economy in which the interactions of producers, households, government, and the rest of the world are mediated via prices and markets. Macroeconomic and resource constraints are respected. Forward and backward linkages and intertemporal adjustments in investment, production, and consumption to changes in economic conditions can be traced. Such models can contain quite detailed information on sectors and households, and so provide something of a "simulation laboratory" for quantitatively illustrating how changes in the economy influence production, trade, employment patterns, and income distributions (Alton et al. 2014).

However, it should also be recognized that CGE models fundamentally are deterministic, providing precise point estimates of the economic outcomes that can be attributed to changes in the economy, but offering limited

1 The dynamic recursive CGE model used for this exercise was developed by Benfica and Thurlow (2017) based on a 2014 social accounting matrix of Malawi's economy (Thurlow 2017).

guidance on uncertainty in those estimates.[2] Consequently, the results from scenarios run through such models, as here, should be viewed as illustrative. If used for policy design, they will need to be validated using other sources of information on the issues examined.

The scenarios run in the economywide model for Malawi involved increasing the annual growth rate in total factor productivity for each of the three sectors, in turn, by 20 percent from current levels.[3] Such increases in total factor productivity growth for Malawi are achievable. Malawi has been underperforming compared with other countries in Africa when measured by labor productivity (value-added per worker) by sector between 2014 and 2019 (World Bank 2020). For the agriculture and industry sectors, Malawi is in the bottom quartile of the 44 countries in Africa south of the Sahara, although for services, it is in the second-lowest quartile.

The outputs of these three simulations are compared with a base, business-as-usual scenario that projects the recent performance of Malawi's economy through to 2030. Differences between the sector-specific results can guide policymakers in allocating resources across economic sectors to better realize particular aspects of the development vision for the country.

The modeling exercise shows that a service sector–led investment plan will lead to the largest returns in terms of overall economic (GDP) and employment growth (Table 6.1). Under the services-led scenario, both annual GDP and employment growth overall increase by around 0.2 percentage points relative to the base case and by 0.1 percentage points relative to the agriculture-led and industry-led scenarios. When compounded annually through 2030, the services-led scenario results in a 4.4 percent larger GDP for Malawi in 2030 over the base case and 4.2 percent more employment. Relative to the agriculture-led and industry-led scenarios, the services-led scenario results in 2030 in a 2.2 percent larger GDP and 2.1 percent more employment.

Increased production of services under the services-led scenario leads to higher demand for food and industrial goods, so both production and employment in the agriculture and industry sectors increase more under the services-led scenario than under the base case. Focusing on growth in the

2 Sources of uncertainty in the use of deterministic models include random variability in the data used to design and operationalize the model, varying accuracy and precision in the model parameters, and possible inaccuracies in the assumptions inherent to the construction of the model (Briggs et al. 2012).

3 Total factor productivity is calculated as the ratio of the value of economic output to the value of production inputs—land, labor, capital, and materials—and is a measure of how efficiently inputs are used in economic production (Benin and Nin-Pratt 2016). It can be calculated for the economy as a whole or, as here, for sectors and subsectors of the economy.

TABLE 6.1 Effect of sector-specific development scenarios on average annual growth in GDP and employment and on gross value-added sectoral shares, 2018 to 2030

	Annual GDP growth 2018–2030, %			
Sector	Base scenario	Agriculture-led	Services-led	Industry-led
Overall	5.5	5.6	5.7	5.6
Agriculture	5.3	5.7	5.4	5.3
Industry	5.7	5.7	5.8	6.0
Service	5.5	5.6	5.8	5.7
	Annual employment growth 2018–2030, %			
Overall	5.2	5.3	5.4	5.3
Agriculture	5.4	5.5	5.6	5.5
Industry	4.2	4.3	4.5	4.1
Service	5.0	5.0	5.0	5.1
	Gross value-added in 2030, % share of total			
Agriculture	28.3	29.2	28.0	28.0
Industry	16.9	16.7	16.9	17.3
Service	54.8	54.1	55.1	54.7

Source: Benson and Hartley (2020), using the Malawi CGE model.

service sector will bring about greater shifts in the structure of the economy of Malawi than would investments in the agriculture or industry sectors, even if accelerated productivity in services will need to be maintained over many years.

But in making decisions around focusing on strengthening Malawi's service sector to propel economic growth and employment, policymakers face an important trade-off across development objectives. If improving welfare and reducing poverty are also central objectives for Malawi's economic development, the agriculture-led scenario achieves the greatest progress toward these aims (Figure 6.1). In part because so many rely on agriculture-based livelihoods, poorer households see appreciably larger improvements in their levels of consumption under the agriculture-led scenario than under the scenarios that concentrate public investments in the other two sectors. The services-led scenario is shown to be relatively effective in providing improved welfare for all, but it is superior to the agriculture-led scenario in this regard only for the wealthiest quintile of households.

FIGURE 6.1 Household welfare changes by consumption quintile under the Malawi CGE model scenarios in 2030, relative to the base scenario

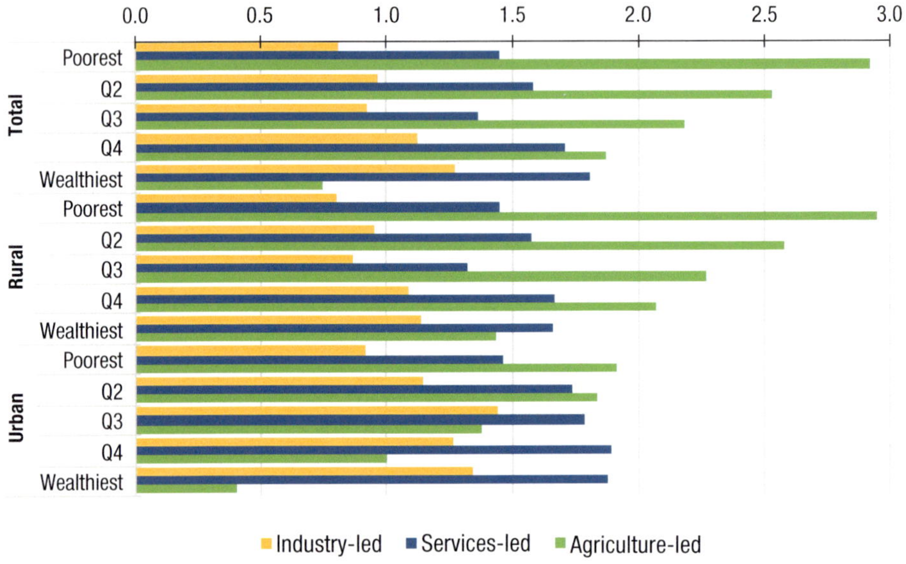

Source: Benson and Hartley (2020).
Note: Q = consumption quintile.

The modeling exercise does not consider how the 20-percent increases in sectoral total factor productivity that were used in the scenarios will be achieved. In the extensive research literature on the topic, the principal drivers of productivity growth commonly recognized are enhanced innovation to create or adopt new productive technologies; greater educational attainment so that workers are more productive; improved market efficiency to enable resources to be more productively allocated within the economy; infrastructure development to enable efficient production; and strengthening of institutions that guide economic policy, regulate economic activities, enforce laws, and safeguard rights (for example, Kim and Loayza 2019).

However, identifying the specific product- or services-focused strategies for the short to medium term that will enable Malawi to achieve the significant increases in output required will be a significant challenge. It will require identifying the products and services for which Malawi has a comparative advantage in production. Based on such assessments, the National Export

Strategy seeks to diversify Malawi's exports into new products and services, while also offering additional support to existing exports that are well positioned to expand their foreign markets (Malawi, MoIT 2012, 2020).[4]

This level of product- and service-specific detail cannot be drawn from the results of these CGE model simulations. Nevertheless, sector-specific guidance for efforts to transform Malawi's economy and sustainably reduce food insecurity can be drawn from the simulation results:

- Significant investment in the service sector will do the most to drive economic growth and increase incomes. Directing investments toward the service sector will provide the greatest increase in incomes in aggregate. Focusing on growth in the service sector will bring about greater shifts in the medium to long term in the structure of Malawi's economy than would investments in the agriculture or industry sectors.
- Accelerated growth in services will not rapidly alter the structure of Malawi's labor force, because increased demand for agricultural and industrial products due to the rising incomes of those working in the service sector will result in a small increase in the share of workers employed in agriculture and industry.
- Moreover, whether the higher aggregate income that results from investing in services will lead to improved access to food for all is not clear. How a services-led strategy will affect food security at the household level will depend on the distribution of the increased income across the population and how food prices change.

4 Three new export clusters were promoted under the National Export Strategy for 2013 to 2018 (Malawi, MoIT 2012)—oilseed and oilseed products, sugarcane and sugar products, and manufacturing. The focus of efforts to promote manufacturing for exports includes manufacturing of beverages, plastics, and packaging; agroprocessing of dairy, maize, wheat, horticulture, and pulses; and prefabrication and assembly of trailers, trucks, agricultural machinery, electronics, and other electrical equipment.

The existing export areas highlighted for receiving continuing attention under the National Export Strategy for 2013 to 2018 were tobacco, mining, tea, tourism, and services, particularly professional legal and accountancy services.

Under the revised National Export Strategy for 2020 to 2025 (Malawi, MoIT 2020), four priority export clusters are identified for promotion:

- Agriculture: legumes (beans, peas, and pulses), edible nuts, oilseed, fruits, hemp, livestock, and fisheries
- Services: tourism, information and communication technologies, creative industries, and professional services
- Manufacturing: value-added processing of tea, oil cake, and cane sugar; plastics packaging; and wood products
- Mining and minerals: uranium, gemstones, coal, and construction materials

- From an equity standpoint, higher investment in services will, at least through 2030, tend to benefit wealthier Malawian households more than poorer ones. Focusing investment in agriculture will result in lower incomes in aggregate than is the case with the services-led scenario. However, significant benefits will still be realized, and they will be disproportionately enjoyed by poorer households. If equity considerations and generating benefits for the poor are important in the short term, an agriculture-led economic development strategy is superior to investments in the other sectors of the economy.
- The industry sector currently is much smaller than either the agriculture or service sectors, so significant increases in productivity growth in industry result in smaller benefits overall than comparable increases in productivity growth in agriculture or services.

There are clear trade-offs in choosing which sector should receive emphasis in any economic development strategy. A services-led strategy will result in higher economic growth overall and broadly rising incomes. However, the poor will not share in the resultant benefits as quickly as wealthier households. In contrast, an agriculture-led approach, although it better meets the needs of the poor and their access to food, may result in continued lagging growth for Malawi's economy, with the result that it does not provide sufficient income and employment opportunities for the bulk of poor Malawian households to exit poverty and realize better lives for themselves and their children.

However, if a primarily agriculture-led strategy of economic development is adopted for Malawi, targeting agricultural productivity enhancement programs to poorer, subsistence-oriented households may not be superior in achieving those aims than focusing such programs on the less poor, commercially oriented smallholders. Benefits to household welfare can come about directly through increased household agricultural production, including subsistence-focused production, but also indirectly through improved flows of income into the household through increased demand from higher productivity households for products and services produced by members of poorer and less productive rural households, as well as through lower prices, such as would result from increased production of food crops with much of the increase sold in the market. All these channels to improve household consumption levels—increased production and incomes, and lower prices—are captured in the way the Malawi CGE model calculates changes in household welfare under different scenarios. This is pertinent to the discussion in Chapter 5 on the leading role that commercially oriented smallholder farmers could play in rural economic development. Targeting such farmers in any agriculture sector–targeted economic development efforts would be particularly effective if Malawi's agricultural markets also were strengthened.

Choosing how best to balance public investments in the various sectors to achieve robust economic growth and transform the structure of the Malawian economy while also meeting the food needs of all Malawians will require a clearer development vision to guide decisions around these complex issues. Trade-offs between development objectives will be inherent to doing so. Nonetheless, the results of this modeling exercise generally support the view advocated by this book that the food security of all Malawians can be achieved through increasing the production by commercially oriented smallholder farming households, strengthening agricultural markets, and increasing nonfarm livelihood opportunities for other rural households.

Broader Investments to Expand Employment Opportunities

At the heart of any discussion on the structural transformation of Malawi's economy is the need to sustainably improve returns to labor. The ambitions that all Malawians have for themselves and their children will be realized principally through their participation in decent work to which they can productively apply their energy, skills, and creativity. Regardless of the development pathway followed, efforts around broad human capital development—improved health and increased knowledge, experience, competence, and creativity—will remain central to propelling economic growth in Malawi. Here we review several broad investment areas that will enable many Malawian workers to obtain more remunerative employment and will accelerate the transformation of Malawi's economy.

Education and health. A healthy and well-trained workforce is needed so that more complex production processes can be used, including in agriculture, and more effective services can be offered to increase the overall output of Malawi's economy. In education, although the vision of reliably providing high-quality free primary education to all children is not yet fully realized, increased focus on technical, vocational, and entrepreneurial training and expanded access to secondary and tertiary education would be valuable. In the health sector, expanded public investments are needed in preventive and curative health services to ensure that children can develop their full economic potential. Well-educated and healthy Malawians expand the human capital stock of the country and its economic potential.

Electricity. Although 42 percent of urban households have access to electricity in Malawi, only 4 percent of rural households do. The population is growing faster than the national electrical system is expanding (World Bank 2019). Power outages are seasonally common during periods of low water flow through the hydroelectric stations on the Shire River. Expanded and more reliable supplies of electricity to Malawi's cities would enable growth in the types of industrial products and services generated there, while also improving the quality of life for poorer urban workers. Expanded rural electrification would also expand agroprocessing possibilities, enlarge local employment prospects, expand local demand for agricultural commodities, help build local wealth, and reduce the use of biomass for the energy needs of rural households, reducing deforestation. Moreover, electrification is important simply for improving the quality of life of rural households. For both economic and human development reasons, massive investment in electrification, particularly in rural communities, makes sense. This could be done for Malawi's urban centers through mains that draw on large-scale domestic generation from Malawi's hydro or coal resources, or by increasingly relying on regional, networked sources of power (Taulo, Gondwe, and Sebitosi 2015). However, particularly for rural households, electricity needs also could be met by propagating local, decentralized mini-grid systems that draw on smaller-scale solar, hydro, or other energy sources (MERA 2017, 2019; Blimpo and Cosgrove-Davies 2019). Many of the decentralized systems, in particular, could be supplied and run by private firms.

Internet connectivity. Increasing investments in Internet connectivity are similarly needed. Trained Malawian workers can be efficient, low-cost providers of Internet-based services of all sorts on the international market (Murray 2017). Malawians can take part in the growing global market for such services. However, a necessary condition for doing so is that fast, reliable, and widespread Internet connectivity be in place across Malawi and skills built among Malawian workers so that they can take advantage of emerging Internet-based demand for such skills. Increasingly, much of this training can be obtained online, adding impetus to improving Internet connectivity across the country in the near term.

Urban development and pro-urban policies. In the past, policy frameworks for development in Malawi have sought to check rural-to-urban migration (Kalipeni 1997) and, in consequence, provided for only limited public investments in urban development and housing in both the major urban centers and smaller towns. However, Malawi's future depends on a structural transformation of the economy with considerable movement of labor and capital out of agriculture and into manufacturing and services. Historically, there are no examples of successful national economic development occurring without an increased concentration in urban areas of economic activities and of the workers involved in those activities. Although the geography of agricultural production is inherently rural, the geography of manufacturing and service provision is strongly urban. Ellis accurately stated that "rural poverty reduction and urban growth are interdependent, and rural poverty reduction requires a much more rapid rural-urban transition than has been occurring.... Future poverty reduction strategies need to be oriented more to increasing people's mobility out of agriculture" (2005, 14–16).

If a successful structural transformation of Malawi's economy is to occur, the social and physical development needed to bring it about will take place in the cities and smaller urban centers of the country as centers of manufacturing and services. Government should work to remove any legal or policy barriers that restrict rural-to-urban migration. This migration will happen, for better or worse.[5] The government must act so that it is for the good, even as it also works to make rural livelihoods more rewarding. This would include government catalyzing significant investment in urban utilities and systems, such as housing, water and sanitation, electricity, transportation, and

5 Two other population movements also have a bearing on the economic development challenges that Malawi faces, particularly in rural areas. Rural-to-rural migration likely will increase in coming decades. Intensifying shortages of agricultural lands in southern Malawi have propelled rural workers and households over the past several decades to central and northern Malawi either to work as seasonal tenants on tobacco estates or to seek land for permanent resettlement (Potts 2006). Given kinship ties, cross-border migration into less densely settled areas of Mozambique, in particular, also is likely to be an option increasingly pursued by rural Malawian households with insufficient land to meet their own needs.

The government, with the support of development partners, has carried out a handful of resettlement programs, particularly through the acquisition of underutilized estate lands that are subdivided and provided to land-constrained rural households from elsewhere in the country. The one such program that was rigorously evaluated found that the beneficiaries saw improvements in their food security (Mueller et al. 2014). However, this was solely due to the households increasing the area that they cultivated, and not to any improvements in productivity. Moreover, if resettlement is in a remote area with poor market access, it was found that such programs are unlikely to enable farm households to be strongly commercially oriented in their production. Enabling the beneficiaries to assure their food security through more than subsistence production alone will require that such programs include local market strengthening.

The other migration stream of potential interest is urban to rural. It is a common pattern for individuals who grew up in rural communities but followed career paths that took them to other areas of Malawi or the wider southern Africa region to retire from those careers by returning to their community of origin. Many of these individuals will establish the basis for their retirement by investing in farming and other businesses in those rural communities well before they retire (Anseeuw et al. 2016). As they do so, these individuals form part of the urban-based commercial farmers subset of households discussed in Box 5.1. However, after retirement and returning to the community, they will then form part of the commercially oriented smallholder farmer category that is at the center of the model of rural economic development discussed in Chapter 5.

communication infrastructure. Urban local governments will need support to effectively manage the development challenges they face through improved technical and institutional capacity and increased financial resources, including improved local taxation systems to finance urban development (Debalen et al. 2017). Malawi's towns and cities need to be developed to handle the rising flows of people who will seek their livelihoods there in coming years. Moreover, making effective investments in urban centers now, so that these towns and cities are attractive locations for industries and for the provision of a wide range of services, will promote further urban investments in the future.

With regard to the role of agriculture in this economic transformation, Malawi's political leaders should be cautious about adopting any "agriculture first" development strategy. That rural households will need continuing support from the government is not in question, and significantly increasing agricultural productivity levels is central to improving the economy of Malawi. Just as there are no examples of successful national economic development without increased urbanization, no principally agrarian country has successfully transformed its economy to being industry- or services-led without first improving its agricultural productivity (for example, Johnston and Mellor 1961; Gollin, Parente, and Rogerson 2002). Government should strongly encourage and support entrepreneurial smallholder farming households both to improve their productivity and to increase the scale of their production.

However, there is a risk that the government, in providing agricultural support to poorer rural households, will reduce their incentives to leave subsistence-based agricultural livelihoods when they would have a better future if they pursue employment off-farm and out of agriculture. These less entrepreneurial or otherwise economically constrained rural households should be offered training opportunities and other incentives to enable them to increasingly apply their labor in other sectors of the economy, where it can be used more beneficially than in low-productivity farming. Many such households may migrate to Malawi's cities, while many who stay in rural areas will find work in nonfarm enterprises that is more profitable than remaining in farming.

Although the transition from farming to nonfarm employment will be difficult for many, the government can assist households to make this transition by providing targeted skills training, building the transport infrastructure, and providing the marketing, communication, and other commercial services required for Malawian workers to succeed outside of agriculture. Assisting households to overcome any shocks to their livelihoods that undermine their food security and their ability to meet the nutritional needs of household members in this transition will require that appropriate social safety net programs be in place. An equally important reason for having such programs of social support is that such shocks may result in the household's reverting to low-productivity agriculture as the basis for its members' welfare. Social programs can assist households that are making such a transition in livelihoods to cope with such shocks so that they can continue to engage in the production and provision of nonagricultural goods and services, and do not forgo the greater economic benefits that they are likely to obtain outside of farming.

Planning for Better Engagement by Malawians in Regional and Global Labor Markets

If, however, a deep and rapid structural transformation of Malawi's economy is not achieved, Malawians will increasingly need to leave the country to achieve their own economic ambitions and to meet the needs of their dependents. Emigration for wage labor, whether temporary or permanent, could be a part of the vision for Malawi's future.

Policy decisions are needed to guide how such labor migration might benefit all concerned—the migrant, his or her family in Malawi, and the country as a whole. For example, greater regional attention to planning and regulating flows of labor will enable Malawian workers to safely find secure jobs in neighboring countries. In addition, low-cost, safe, formal channels for sending funds will be of significant benefit to both the migrant and those being supported by the remittances. More broadly, Malawians in the diaspora need to be seen as a longer-term resource that can make valuable contributions toward the achievement of the country's development goals financially, by transferring skills, and by introducing and championing innovations to broaden the set of options considered in prioritizing and mapping development pathways for the country. A first step in creating strategies to ensure that regional and international labor emigration contributes to achieving the development vision of the country is to assemble much better data than currently can be found on the numbers of Malawians working outside of the country, the sort of work they are doing, and the remittances they are sending back (IOM 2015).

Emigration for wage labor, whether temporary or permanent, has been an important economic strategy for many Malawians since early in the colonial period. Although (mainly undocumented) wage labor migration from Malawi will continue and likely intensify, particularly to South Africa, we should also expect to see increasing informal movement of smallholder farmers from

Malawi into neighboring Mozambique as they seek to derive their subsistence from the less densely settled arable areas there. Similarly, many of the most well-trained Malawians will pursue their careers outside of the country—the aphorism from decades back that "there are more Malawian physicians in Manchester than in Malawi," though perhaps no longer quite as true as it once was, still resonates (Masanjala 2018; Bhoojedhur and Isbell 2019). The conditions of service and salaries for Malawian professionals are significantly poorer in Malawi than in South Africa, Europe, or other higher-income regions where the skills of those professionals are in demand.

If such a vision for the economic future of Malawi is not desired, then Malawi's leaders will need to undertake the investments and other actions necessary to ensure that those Malawians who might otherwise migrate can achieve their own economic ambitions as satisfactorily in Malawi as outside the country. This may involve as an interim measure targeted incentive schemes to retain well-trained Malawians within the domestic labor force. However, in the absence of such actions, wage labor migration will occur as the default option for any economically ambitious Malawian worker.

The broad and long-term perspective on Malawi's economic development outlined in this chapter goes beyond what most would view as central to achieving food security in the country. However, short-term technical solutions will not enable Malawi to sustainably achieve food security. For many Malawians, their interests, including both their food security and the nutritional well-being of their children, will be best served by leaving agriculture and working in the industry and service sectors. This will not be a costless process socially—there will be sectors of Malawi's population that, at least in the short term, will find that the livelihoods upon which they relied are no longer sufficient to reliably meet their basic needs, including food. These individuals and households will benefit from targeted assistance that facilitates their seizing some of the new livelihood opportunities that emerge as the structure of the economy changes, rather than being left behind in poverty. Structural transformation of Malawi's economy must be a central element in any longer-term human development vision for the country.

Chapter 7

ACTIONS TO ADVANCE TOWARD A RELIABLY FOOD-SECURE MALAWI

This closing chapter summarizes a set of policy and public investment recommendations that were discussed in earlier chapters:

- Continue efforts to increase agricultural productivity and reduce volatility in food supplies.
- Strengthen agriculture and food markets in Malawi by ensuring predictable government engagement, adopting policy stances that are supportive of agricultural market traders, and expanding participation in regional markets.
- Target agricultural development investments to commercially oriented smallholder farming households, while also addressing the needs of all households that are food insecure.
- Develop incentives for Malawians to work outside of agriculture, ensuring they have the skills to productively do so.

Attention is also given to reforming the institutional framework through which public efforts are made to address food insecurity in Malawi. Food security policy should be informed by a broader conceptualization of how Malawian households can reliably access the food they require beyond subsistence production. Institutionally, an approach to food security is needed that goes beyond agriculture alone to encompass multiple sectors and draws upon high-level leadership, coordination, and accountability.

As was highlighted in the introduction, this book takes the position that the dominant subsistence orientation of most farmers in Malawi will never sustainably assure the country's food security nor provide the economic growth necessary to propel a structural transformation of the economy. To achieve a food secure Malawi will require economic development policies and programs and public investments that lay the foundation and provide incentives for Malawians to improve their well-being through economic pursuits outside of subsistence agriculture. In parallel, public investments must be provided to improve the productivity of those Malawians who will continue to farm. A critical component for achieving these changes will be to significantly improve the functioning of domestic food and agriculture markets to increasingly allow individual Malawian households, including farm households, to produce less of, and rely on markets for more of, the food they eat.

How best to address food insecurity and to overcome the factors that constrain the development of the agriculture sector and rural economies are questions that are not unique to Malawi. Similar development challenges face most of Malawi's neighbors and other low-income countries across Africa and elsewhere. What are the lessons from this examination of the challenges faced by Malawi in assuring food security and promoting agricultural and rural economic development that can be adapted by these countries to employ in their own specific contexts? Here we draw out some that are broadly applicable both to Malawi and to many other low-income countries.

Inherent to the development approach advocated in this book is weakening the agricultural identity of food security. Conceptually, food security is not synonymous with household own production or, nationally, with domestic agriculture. There are a broad range of mechanisms beyond own production through which households can obtain the food that they require. Closely associating food security with agriculture ignores many food insecurity problems, including those of nonproductive rural and urban households. The food security of the most vulnerable households likely can be best addressed through specific policies and actions on social protection. In contrast, the food security of productive households likely can best be ensured through sustainable and broad economic development, giving households, through dynamic and efficient labor and product markets, a broader range of options for meeting their own food needs. Moreover, while national self-sufficiency in food production may be justified on economic and risk management grounds, it should not be accepted as an unexamined doctrine of national policy. Agriculture is a necessary means, but neither a sufficient nor an optimal means on its own, to assure household and national food security.

Raise Agricultural Productivity

Continued and expanded efforts in agricultural research and in providing agricultural extension services will enable more efficient use to be made of productive resources for agriculture. Obtaining more output per unit of land, labor, or other input used remains an important agricultural development goal. We discuss below two issues specific to Malawi, but not exclusive to the country, on which its government should act to foster increased agricultural production—first, a range of efforts to reduce volatility in maize production and maize prices, and second, input subsidy reform.

Variability in aggregate maize production from year to year results in significant supply-induced inconsistency in seasonal maize price patterns in domestic markets in Malawi. The uncertain prices that households face in acquiring their staple food in the market convince many that the potential risks to the welfare of their household associated with relying on the market for food are too high. So most households remain engaged in subsistence agricultural production, selling in the market only the limited amounts of grain from their plots that exceed their household needs. These interlinked crop production and market performance factors result in weak incentives for farming households to produce surpluses of maize and, consequently, in the thin markets for maize in Malawi. It remains uncertain and potentially risky for most Malawians to rely primarily on the market for their access to food. Public investments in efforts to enhance maize production, if made alongside complementary efforts to reduce transaction costs and to limit uncertainty in government interventions in maize markets, will stabilize and strengthen those markets by increasing producer incentives to provide more consistent supplies of maize. Larger and more reliable supplies are a necessary, even if not sufficient, element in deepening maize markets in Malawi.

There is a range of technical approaches that would sustainably raise production. Adaptive agricultural research is critical to doing so. Efforts are underway to develop more water-efficient maize varieties that are less susceptible to yield loss under erratic rainfall conditions. Research to address pests, diseases, and other biotic constraints to crop production needs to be maintained and intensified. Continued efforts to refine crop and soil fertility management practices are required so that farm households consistently realize much higher yields. Expanded irrigation is part of these efforts. Although staple food crop production under irrigation is unlikely to be financially viable, the income benefits for producers and the dietary benefits for consumers of increased irrigated production of high-value crops, particularly of nutrient-dense vegetables and fruits, can lend greater stability to food markets and improve diets. For all these technical approaches to improved agricultural production, effective agricultural extension services are a necessary corollary, whether provided by government, nongovernmental organizations, or private firms. Information on how to realize greater crop productivity should flow to farm households using any effective means, whether in person or, more remotely, using new communication technologies.

The potential contributions that larger commercially oriented producers could make to expanding national food production and stabilizing food markets in Malawi require further examination. Such efforts should determine how best to encourage commercially oriented smallholders who are able to scale up their production to do so. However, addressing factors that limit the participation of the so-called agricultural estate subsector in commercial food crop production is also required. The identification of clear economies of scale for food crops, possibly under different production practices than are currently employed, if profitable, could significantly improve national food security. Further economic and agronomic research on this topic is needed. However, critical to improving incentives for such farmers to expand their commercial production of food crops is for the government to be consistent in the level of oversight it has over food production and marketing, and in specifying under what food (in)security conditions it will intervene and how it will act to improve the country's food situation. Many of the restrictions that the government imposes and arbitrary interventions it makes in maize pricing, marketing, and trade should be removed. These result in almost absolute disincentives for larger crop producers to engage in commercial maize production, which increases the probability of food insecurity crises and restricts the options the country has for dealing with them when they arise.

Pursuing avenues to ensure that farmers have access to modern inputs without a subsidy program is required if the production benefits of the use of such inputs are to be sustained. Raising maize prices and, at the same time, identifying ways to provide inorganic fertilizer (particularly nitrogen) at lower cost to Malawian farmers are the principal elements of any reforms that would enable the sustainable use of inorganic fertilizer to improve crop production. There are significant opportunity costs in the heavy government financing of subsidized fertilizer, which leaves little funding for other public investments in the agriculture sector. Some of these missing investments, such as those in agricultural research and extension, would increase the efficiency with which inputs are used, increase production, enhance farm household incomes, and improve access to food for many more households. Similar arguments can be made for reallocating some of the funding now used to subsidize agricultural inputs to other public investments, including market information systems and transportation and communication infrastructure, that would provide farmers with greater returns on the crops they produce and, hence, greater incentives to increase production. However, such policy reforms, particularly if they require higher food prices, will face significant political headwinds.

Food Security Better Achieved through Stronger Markets Than from One's Own Fields

Markets are becoming a primary source of the food that Malawian households consume. As households increasingly are unable to meet their food needs solely from their own farm production and as workers consequently more and more seek livelihoods outside of farming, the role of markets as the principal source of food will intensify. Food security strategies should reflect the entire food system and seek to strengthen the various pathways through which households obtain food. Privileging subsistence production and, consequently, promoting primarily agricultural solutions to assure the food security of the country reflects a misdiagnosis of both the challenges and the opportunities Malawian households encounter in meeting their food needs. In this, the Malawian situation is not very different from that seen in food systems in many other low-income countries.

Because agricultural and food markets have been consistently undermined over recent decades, intensive efforts to strengthen those markets should be a key part of any effort directed to sustainable food security in Malawi. These efforts include putting in place policies and programs that are supportive of agricultural market traders, more closely specifying the conditions under which the government will intervene in markets and how it will do so, and building the business capacity of farmers and farmer groups so that they engage in the marketing of their produce from a position of knowledge. Government has a legitimate role in regulating markets to ensure transparency and fair trade, to sanction dishonest behaviors, to uphold quality standards that affect prices and food safety, and to strengthen access for all market actors to price and other market information, among others. However, it should not exercise this role in an uninformed and haphazard manner—predictability is important as to the market conditions under which government will intervene and impose any heightened restrictions on marketing and trade. If these and other market-strengthening efforts are successfully implemented, many more people will have reliable access to sufficient nutritious food.

Moving ahead with these efforts will require a change in the policy stance of food security stakeholders—they will need to recognize traders as central actors in ensuring food security in the country. Traders mobilize their own resources to obtain food and other products from farmers and supply it to consumers and processors. However, with little justification, politicians generally see traders as social parasites who work against the best interests of both farmers and consumers. This view should be challenged. Although traders are certainly self-interested, just as are both the producers who sell to the traders and the consumers who buy from them, it is not in their longer-term interest to operate in a manner that is unfair or is not determined by market conditions. Within a competitive market, it is irrational for traders to take advantage of those with whom they are repeatedly transacting their business—they will not stay in business long. As a matter of policy, the activities of traders should not be undermined. Rather, traders should be supported in their role to ensure that the food available within Malawi and the region is provided to those who need it in a sustainable and beneficial manner for all concerned. In general, agricultural markets in the country must become sufficiently inclusive and efficient to foster stronger commercial engagement by producers, processors, traders, and consumers, both to improve livelihoods and to ensure reliable access to food.

Similarly, the government can support producers to help them engage more profitably in markets. One of the principal arguments for the formation of farmer groups and cooperatives is the benefits that individual farmers can obtain by marketing their produce jointly with others, obtaining a higher unit price from traders who seek to buy in large quantities. The government can aid farmer groups by building the abilities of their leaders and providing the information they require to negotiate better prices for their members. State agencies also have an important regulatory function to play in overseeing the contract farming arrangements that are increasingly being used in Malawi, particularly in tobacco production. These arrangements can be an important means to increase the engagement of smallholder farmers in specialized, commercially oriented production and to build their farming and marketing skills through doing so. But this will happen only if the contracts are mutually beneficial, with a fair sharing between contracted producers and buyers of the risks of production and marketing—something that prudent government oversight can play a part in assuring.

Beyond strengthening domestic markets to build food security, Malawi also should significantly expand its role in regional food systems. Regional food trade provides significant economic opportunities for farmers and makes all of the national food systems involved more secure and diverse. Malawi's trading partners will welcome access to Malawi's farm produce to supplement what their own farmers produce. Equally, Malawian food consumers are better served by having access to the much larger supplies of food available regionally, giving them more reliable and lower-cost access to food than if they rely solely on the production of Malawi's farmers.

However, reliance on regional and international markets for food, although enabling a country to better manage shocks to domestic food crop production, does entail risks related to the concurrence regionally of national food deficits with adverse regional food price movements. In Malawi, the strategic grain reserve managed by the National Food Reserve Agency will continue to be an important means for the country to manage, at least in the short term, some of the risks associated with any breakdown in regional supplies of food.

The future of ADMARC, in contrast, is less clear. We take the view that the social costs of the parastatal institution in recent decades have far outweighed any benefits it has provided to farmers or consumers. It is not in the long-term interest of the development of Malawi's agriculture sector to have the parastatal engage in cash-crop value chains that are reasonably profitable for the private firms and producers participating in them. Although it may be financially attractive to allow ADMARC to participate in such value chains to finance the costs of its engagement in food markets, its primary role should be as part of Malawi's food system and not part of broad agricultural development and marketing, despite its name. Any profit orientation on the part of ADMARC should be challenged. Consequently, if ADMARC is to be retained as a mechanism through which the Malawian government intervenes in food markets, the government will need to provide sufficient financing and professional management, in keeping with the importance of ADMARC's role within the country's food system.

Given the inherent political considerations, however, it is unlikely that ADMARC reforms will result from technocratic assessments alone. Both civil society and the private sector have a stake in reforming ADMARC and will need to engage in advocacy efforts. Civil society's continued advocacy for increased accountability around the substantial government subventions ADMARC receives is valuable. The private sector rightly should protest the unfair competitive position of ADMARC in Malawi's agricultural markets, given the government resources regularly placed at its disposal. Both sets of stakeholders should also ensure that any positive functions that ADMARC now performs to improve the food security of vulnerable Malawian households continue to be supported in any reform process, whether by the parastatal itself or through alternative mechanisms.

Finally, the government has an obligation to provide the public services needed for effective market operations that the private sector cannot be expected to provide. Throughout the post-independence period, the government of Malawi has invested considerable resources in building roads and strengthening other transportation and communication services. Further investment and deeper penetration into rural areas continue to be required. Farmers in remote areas will not get a chance to participate in markets if it makes no business sense for traders to engage with them due to the prohibitive costs of communication and transport.

Equally important are improvements in the quality of information that is provided—especially on crop production and market prices—and in its dissemination to all market participants. If both parties to a sales transaction are using dependable information on prevailing commodity prices to guide their decision to engage in that transaction, the risk that the transaction will fail is significantly reduced. Similarly, if the government is considerably more confident in its estimate of the amount of maize available in the country at any time, it is more likely to act to address any concerns around food insecurity in a manner that is supportive of existing, private sector-organized market mechanisms that can ensure access to food for those who need it. Without reliable information on maize production and stocks, the government will respond to perceived food crises in an uninformed manner that, as has been shown in the past, is likely to exacerbate such crises.

Adopt a Differentiated View of Rural Communities to Foster Economic Transformation

Commercially oriented smallholder farming households should be at the heart of efforts seeking to foster rural economic growth led by the country's agriculture sector as a path to broader economic transformation. Focusing public investment on such households to transform the rural economy will, over time, bring the benefits of these investments to a much larger share of rural households than would be the case under alternative approaches. Commercially oriented smallholder farming households are uniquely positioned within rural communities to serve as engines of rural economic growth and to meet the food needs of their communities. As their farm production expands and their incomes increase, they will demand more of the labor-intensive and generally nonmarketable (outside of the local area) goods and services that the many rural asset-poor households produce—construction, transport, education, health, and other social services; furniture- and handicraft-making; food and beverage processing; and the like. There are significant positive spillover and multiplier effects in rural communities from the economic success of these commercially oriented smallholder farming households and the increased demand for local goods and services that their success generates.

This approach to rural economic transformation is developed through a consumption pathway—increased demand by commercially oriented smallholders for the nontradables that other economically active rural households produce locally. Larger-scale farmers, whether foreign investors or citizens, do not provide these broad benefits to local rural economies, inasmuch as they tend not to reinvest the income from their farming in the local community through increased consumption of local products. Although these larger farmers may be important for the local diffusion of more productive technologies and for improving the prospects of commercial agricultural production for all local farmers, their local consumption patterns will not contribute much to broader local employment. Similarly, poor, subsistence-oriented households that engage in some farming also do not contribute significantly to this process of economic transformation in rural communities, because they sell very little of their production and generate little, if any, new wealth in their communities through their agricultural activities. Recognizing economic differentiation in rural communities across the country and targeting agricultural support to commercial smallholder farming households will raise crop productivity overall, accelerate the economic growth of rural communities, and provide poorer, subsistence-oriented households with more remunerative means to obtain their livelihoods than from farming.

But at the same time as efforts are made to improve the economic returns realized by commercial smallholder farming households, the economic development of other rural households will need significant support. These households, which typically engage in some subsistence farming alongside other economic pursuits, make up the largest share of households in rural communities across Malawi. For these other rural households, pursuing livelihoods outside of agriculture will better serve their economic prospects, improve their well-being, and contribute to the economic growth of their communities. The government and its partners should work with these households to build their capacity and create opportunities outside of farming so that they will increasingly find that their food security and overall livelihoods are best ensured through engaging in more specialized work. This can be done by providing the programs and incentives to enable workers in these households to build their own abilities to successfully and profitably operate nonfarm enterprises or to find jobs that provide a good wage. Through its investments in public goods and appropriate regulations, the government can establish a broader economic environment that is more conducive to enabling these households to improve their livelihoods and incomes from nonfarm activities, whether in rural communities or elsewhere.

The government also has a continuing duty to ensure that the food needs of these other rural households are met. Their food security is at considerably more risk to production shocks than is that of commercially oriented smallholder farming households. Continued government support to such households is required to help make their farms as productive as possible. For the aims of both expanding the production of commercially oriented smallholder farming households and, in the short to medium term, increasing the food security of poor rural households, the policy and strategic plan of the government around agriculture—the NAP (Malawi, MoAIWD 2016b) and the NAIP (Malawi, MoAIWD 2018)—are appropriate. The NAIP will need to be sufficiently funded and professionally implemented so that it meets many of the needs of all farmers.

To best support rural transformation, NAIP implementation should explicitly target commercially oriented smallholder farming households under some of the interventions making up the agricultural investment plan, while others should benefit all farming households. Those NAIP programs that are focused on food security and improving the quality of the diets of rural households should be universal. All those who farm to produce some of the food they consume—both nonpoor, commercially oriented smallholder farming households, and poor, subsistence-oriented households that engage in some farming—should benefit from such programs. But for NAIP programs that seek to expand the participation of smallholder farming households in agricultural markets or those for which commercial viability is requisite to the success of the program, targeted approaches are needed. Market-centered strategies will have the greatest impact if they are principally targeted to commercially oriented smallholder farming households. Prudent targeting of the different government programs in the agriculture sector is needed to better reflect the sharply different economic development and agricultural transformation potential among households that now farm in Malawi.

Agricultural strategies, however, cannot put an end to the food security challenges of households that are not economically productive. Here two groups are of concern—first, the chronically poor who are not able to work and who may be unable to fully rely upon others in their family or community to meet their dietary needs, and second, those who have experienced a significant adverse shock to their livelihoods and therefore experience significant food insecurity for a period of time as they work to rebuild those livelihoods. The food security of both types of vulnerable households is best addressed directly through specific social protection support. Agricultural strategies will certainly be critical over the medium term for assisting farming households that have suffered a severe livelihood shock. However, meeting the food security requirements of all will not be achieved without widespread and sufficiently funded programs of support targeted toward those who are not economically productive, whether chronically or temporarily. Social protection, both traditional forms based in the community and programs motivated or provided by government, is an important component of food security in any country. Governments have a responsibility to ensure that such support is provided to those in need.

In sum, a significantly more diverse and specialized, but also more integrated and mutually dependent economy should be the goal of the government of Malawi. In such an economy, particularly in rural communities, agriculture will continue to play an important, but not an exclusive or pervasive, role. In this way, food security will follow. At the same time, directly addressing the dietary needs of those households and individuals who are facing either chronic or temporary food insecurity will be part of a comprehensive approach to ensuring that all people have reliable access to food.

Structural Transformation

The efforts of Malawi's leaders over the longer term increasingly should be directed to transforming the country's economy and supporting a reorientation in the livelihoods its citizens pursue from reliance on subsistence agriculture to increased engagement in the service and industry sectors. This is a farsighted and long-term aim, but it is necessary if a vision of a country with no hunger or malnutrition and a healthy, productive, and creative population is to be achieved in coming decades. Even as the agriculture sector is transformed from being driven primarily by the own-consumption production decisions of smallholder farming households to a more specialized and commercial production orientation, so too the economy as a whole must be transformed so that increasingly all workers are able to use their skills and talents in more productive ways than they currently can.

Sustainable improvements in labor productivity must be at the center of any strategy for structural transformation of the economy. The ambitions that all people have for themselves and their children will be realized principally through their participation in decent work to which they can productively apply their energy, skills, and creativity. To transform the economy, a healthy and well-trained workforce is needed to expand the human capital stock of the country and its economic potential. In addition to building their human potential, all workers need access to tools, resources, and public services that will allow them to realize their economic potential. This includes ensuring that all can benefit from electricity in their work and lives. Similarly, increasing investments in Internet connectivity are needed to build digital proficiencies and let workers take advantage of high global demand for such Internet-based skills.

Successful structural transformation of the national economy will principally take place in the cities and smaller urban centers of the country that are the hubs of manufacturing, services, and trade, including trade of agricultural commodities. Efforts and policies to reduce rural-to-urban migration and to give precedence to agriculture-sector development in national economic development plans are not in the interests of the country and its economy in the medium to long term. Prudent investments are needed in urban systems, housing, and transportation and communication infrastructure to accelerate urban-based economic development and the flow of labor to urban manufacturing and services jobs. Expansion of urban utilities and systems, such as housing, water and sanitation, electricity, and transportation and communication infrastructure, should be prioritized. Malawi's towns and cities need to be developed now in order to handle the growing flows of people who will seek their livelihoods there in coming years. Moreover, increased government attention in the near term to the development of urban services and infrastructure will make those towns and cities attractive locations for additional private investment. Thereby, a much broader set of resources can be leveraged for their further development.

Food Security Policy and Institutional Reform

Malawi currently does not have a clear or sufficiently expansive food security policy. This is because food security is seen principally as an outcome of sufficient domestic agricultural production. Self-sufficiency in food production is viewed as an ideal to attain both at household and at national levels. It is not. Food security policy in the country should be informed by a broader conceptualization of how households can reliably access the food they require. A more expansive policy should include making the investments and providing the incentives that will result in more efficient food crop production for both own consumption and sale. It should also give equal, if not greater, weight to the understanding that household food security in the longer term is better ensured through increased income from both farm and, increasingly, off-farm sources to purchase food in much more vibrant markets than are presently found in the country. A greater market orientation to both crop production and food security will increasingly allow farmers to focus their efforts on the crops they can best produce and rely on the market to obtain those foods they need that are more efficiently produced by other farmers. It also will enable far more households to work outside of agriculture altogether, pursuing considerably more remunerative economic activities, confident that they always will be able to obtain the food they require from the market.

Leadership for creating the conditions to reliably ensure food security for all Malawians must come from a government institution with a more encompassing view on how the country's citizens could access the food they require. As long as the Ministry of Agriculture has primary responsibility for the country's food security, Malawi will continue its unsuccessful efforts to meet most of its food needs through subsistence agricultural production. A broader vision of food security, encompassing multiple sectors and drawing upon high-level leadership and coordination, is required. As both domestic and international markets are important complementary components to food production for making food available to all, both the Ministry of Industry and Trade and representatives of the private sector involved in food processing and marketing need to be closely involved in setting food security policy.

Food security is a critical dimension of national security and economic resilience. It is not a sectoral issue. For this reason, leadership would best come from a cross-cutting oversight agency of government at the level of the Office of the President and Cabinet or Office of the Vice President. Clear vertical coordination mechanisms would be required in such an institutional architecture to ensure that local leaders at district and subdistrict levels, who understand local food systems better than do national leaders, are given the information they need and the authority to guide local implementation of food security policies and strategies. The Ministry of Agriculture would remain a central technical agency contributing to the country's food security. However, the strategic policy orientation of the country in this regard needs to expand beyond viewing food security solely through the lens of self-sufficient and subsistence-focused agricultural production. A multisectoral approach will establish the policies and strategies needed to reliably achieve food security at both national and local levels.

Finally, the government should set for itself an ambitious vision for how food security is to be achieved for all households and individuals nationally. In the coming decades, a market-centered food system needs to increasingly supplant the current food system based on self-sufficient production. Higher productivity achieved by more specialized, commercially oriented producers of food will allow more Malawians to rely on the market for the food they consume at reasonable prices, with an increased reliance on social cash transfer mechanisms to meet the welfare needs of the most vulnerable. However, we should be clear that the transition to a market-centered food system will take time to complete, will involve some form of both agricultural input subsidies and ADMARC for some time to come, and will require continued government oversight even when achieved.

APPENDIX TO CHAPTER 5

The second subsection of Chapter 5, titled "Building a Rural Development Strategy for Malawi around Commercially Oriented Smallholder Farmers: An Empirical Appraisal," presents a profile of the households in each of four categories of households derived from a conceptual typology of Malawian households formulated to guide strategic economic development. This profile was developed using nationally representative data from the Malawi IHS4 conducted over 13 months in 2016/17. Chapter 5 presents tabulations on the characteristics of households in each category that are most salient to agricultural and rural economic development in Malawi. Additional tables in this appendix present data on the participation of these households in credit markets and in government development and social safety net programs. A third table explores the economic shocks that were most commonly experienced in recent years by households in each category, with particular attention to market-related shocks.

Access to credit. Table 5A.1 provides information on participation in credit markets by the households in the four categories. Fewer than a quarter of households reported receiving any credit, whether formal or informal, in the previous year. Commercially oriented smallholders are more likely than other rural households to have obtained a loan. Although they are no more likely to have obtained those funds from a formal institutional lender than are other households, their reasons for obtaining the loan are quite different from those of other households. The loans obtained by commercially oriented smallholders were much more likely to be for agricultural purposes and much less likely for consumption than were the loans obtained by other types of households.

Participation in government development and social safety net programs. To understand how the government targets some of its development programs and responds to the chronic economic challenges of those

TABLE 5A.1 Credit use of households in the different economic categories, 2016/17

Characteristic	All households	Commercially oriented smallholder farmers	Other productive rural households	Not economically productive	Urban households
Obtained credit in past year, % of households	23.1	28.5 ***	22.2	15.4	28.4
Credit from an institutional lender, % of those obtaining credit	44.9	44.5	46.1	41.9	42.5
Credit for agricultural purposes, % of those obtaining credit	17.2	32.6 ***	19.4	20.0	5.9
Credit for consumption purposes, % of those obtaining credit	51.1	35.4 ***	53.4	66.8	45.3
Observations	*12,447*	*636*	*8,412*	*1,161*	*2,238*

Source: Author's weighted analysis of 2016/17 Malawi Integrated Household Survey 4 (Malawi, NSO 2017).

Notes: Asterisks on the estimates for "commercially oriented smallholder farmers" indicate the statistical significance of the results of a Wald test on differences in estimates between households in this category and those in the "other productive rural households" category. * = $p < 0.10$, ** = $p < 0.05$, *** = $p < 0.01$.

households that are not economically productive and the occasional acute economic shocks faced by other households, Table 5A.2 examines participation in these programs by each category of households. The FISP is the government's principal agriculture-sector development activity. Since its launch in 2005/06, the program has not been consistently targeted—it is not clear whether it is a production-focused program that seeks to improve access to fertilizer and improved seed for the most productive smallholders or, in contrast, a safety net program to improve the food security and incomes of poorer rural households. Analysis of the IHS4 (Malawi, NSO 2017) shows that commercially oriented smallholders are more likely than other rural households to be beneficiaries of FISP, suggesting that FISP in its implementation has a stronger focus on production than on safety net support.

The IHS4 questionnaire also asked whether members of the survey household benefited from any of a range of other social safety net–type programs. Households in the "not economically productive" category were the most likely to have benefited from such programs, an encouraging result. Moreover, members of other productive rural households were more likely

TABLE 5A.2 **Participation in social safety net programs in past year by households in the different economic categories, 2016/17**

Characteristic	All households	Commercially oriented smallholder farmers	Other productive rural households	Not economically productive	Urban households
Farm Input Subsidy Program (FISP) beneficiary, % of households	17.2	24.9**	19.9	16.2	6.1
Safety net programs, households with a member who was a beneficiary in past year, %	40.2	36.9**	43.8	51.6	22.6
Estimated value per capita of assistance received by households receiving assistance in past year, MK	10,177	11,864	11,271	6,338	6,170
Food assistance beneficiary, % of households	55.4	50.9**	61.5	53.7	17.6
Northern region	23.9	22.1	26.3	15.0	21.4
Central region	53.9	50.6	58.5	52.9	15.7
Southern region	61.6	60.4	68.1	58.7	16.9
Public works program beneficiary, % of households	20.7	23.5	20.0	16.1	29.3
Nutritional support program beneficiary, % of households	31.9	24.7	28.7	41.7	46.0
Cash transfer program beneficiary, % of households	11.5	15.3	12.4	12.2	3.0
Observations	*12,447*	*636*	*8,412*	*1,161*	*2,238*

Source: Author's weighted analysis of 2016/17 Malawi Integrated Household Survey 4 (Malawi, NSO 2017).

Notes: Asterisks on the estimates for "commercially oriented smallholder farmers" indicate the statistical significance of the results of a Wald test on differences in estimates between households in this category and those in the "other productive rural households" category. * = $p < 0.10$, ** = $p < 0.05$, *** = $p < 0.01$. MK = Malawi kwacha; in late 2016, US$1 = MK 715.

to have participated in such programs than those in commercially oriented rural households. However, although the estimated per capita level of benefits received by participating households is roughly equivalent between the households in the productive rural categories, households in these categories received significantly greater benefits than did members of households categorized as being not economically productive. This is a negative finding because the "not economically productive" households likely are the preferred targets for most of these programs.

As has been noted, the 2016/17 IHS4 was carried out during a food insecurity crisis caused by drought conditions during the prior cropping season, particularly in the Southern region, but also in the Central region. Over much of the survey period, the 2016/17 Food Insecurity Response Program was being implemented to provide humanitarian assistance to vulnerable households. Information on whether a household was a beneficiary of any food assistance (not exclusively, but primarily from the Food Insecurity Response Program) was collected in the IHS4. Table 5A.2 shows, at the national level, that commercially oriented smallholder households were somewhat less likely to have been beneficiaries of food assistance programs than other productive rural households. But unfortunately, we do not see evidence of preferential targeting of this assistance to households in the "not economically productive" category.

Participation in public works, nutritional support, and cash transfer programs is also considered in Table 5A.2. The targeting criteria for these programs likely varies—public works programs being targeted to households with sufficient labor, nutrition programs to all households with young children, and cash transfer programs to those not economically productive. Although no differences in participation in these programs are seen between the two productive rural household categories, there are some sharper differences in participation for the other categories of households. However, we do not see that households that are not economically productive are preferentially targeted by cash transfer programs.

Types of economic shocks experienced. Many of these government programs are designed to help households cope with adverse economic shocks and to increase their resilience. The most significant shocks that rural households in Malawi experience are primarily covariate in nature, affecting many or all members of the community at the same time and stemming from a change in weather or market conditions, or pest and disease outbreaks (Table 5A.3). Although covariate shocks are also the most common sort of economic shocks experienced by urban households, it is considerably more common for urban households to feel that the most important recent shock that they experienced was idiosyncratic, affecting their household alone and stemming from a birth, death, or illness in the household, or a sharp change in household income. We also see that price-related shocks were more commonly reported by urban

TABLE 5A.3 **Most important shock to household welfare in past year for households in the different economic categories, 2016/17**

Characteristic	All households	Commercially oriented smallholder farmers	Other productive rural households	Not economically productive	Urban households
Idiosyncratic shock experience—most significant shock to household welfare in past year affected only the household, % of households	10.9	6.8	8.1	6.5	24.4
Covariate shock experience—most significant shock also affected many other households, % of households	89.0	93.2	91.8	93.5	75.4
Price-related shock—most significant shock was price-related, % of households	32.8	35.6***	27.4	27.9	53.8
Low crop price was the price shock, % of households experiencing a price-related shock	3.6	9.3*	4.7	3.0	0.7
High farm input price was the price shock, % of households experiencing a price-related shock	30.6	55.2***	39.6	36.2	8.2
High food price was the price shock, % of households experiencing a price-related shock	65.7	35.5***	55.7	60.7	91.1
Observations	*12,447*	*636*	*8,412*	*1,161*	*2,238*

Source: Author's weighted analysis of 2016/17 Malawi Integrated Household Survey 4 (Malawi, NSO 2017).

Notes: Asterisks on the estimates for "commercially oriented smallholder farmers" indicate the statistical significance of the results of a Wald test on differences in estimates between households in this category and those in the "other productive rural households" category. * = $p < 0.10$, ** = $p < 0.05$, *** = $p < 0.01$.

households than other sorts of shocks, reflecting their greater economic dependence on the market.

Of the covariate shocks reported by the IHS4 sample households, shocks related to rainfall patterns were most common. However, many households also reported price-related shocks. The type of price shock reported varied by household category. Commercially oriented smallholder households were more likely than other households to report production-related price shocks—high input or low crop output prices. Households in the other categories were more likely to report high food prices as an important price-related economic shock experienced recently.

REFERENCES

Abate, T., M. Fisher, T. Abdoulaye, G. T. Kassie, R. Lunduka, P. Marenya, and W. Asnake. 2017. "Characteristics of Maize Cultivars in Africa: How Modern Are They and How Many Do Smallholder Farmers Grow?" Agriculture and Food Security 6:30.

Adu-Gyamfi, J. J., F. A. Myaka, W. D. Sakala, R. Odgaard, J. M. Vesterager, and H. Høgh-Jensen. 2007. "Biological Nitrogen Fixation and Nitrogen and Phosphorus Budgets in Farmer-Managed Intercrops of Maize–Pigeonpea in Semi-arid Southern and Eastern Africa." Plant and Soil 295:127–136.

Akinnifesi, F. K., P. W. Chirwa, O. C. Ajayi, G. Sileshi, P. Matakala, F. R. Kwesiga, H. Harawa, and W. Makumba. 2008. "Contributions of Agroforestry Research to Livelihood of Smallholder Farmers in Southern Africa: 1. Taking Stock of the Adaptation, Adoption and Impact of Fertilizer Tree Options." Agricultural Journal 3 (1): 58–75.

Allan, W. 1965. The African Husbandman. Edinburgh: Oliver and Boyd.

Alton, T., C. Arndt, R. Davies, F. Hartley, K. Makrelov, J. Thurlow, and D. Ubogu. 2014. "Introducing Carbon Taxes in South Africa." Applied Energy 116: 334–354.

Alwang, J., and P. B. Siegel. 1999. "Labour Shortages on Small Landholdings in Malawi: Implications for Policy Reforms." World Development 27 (8): 1461–1475.

Anseeuw, W., T. S. Jayne, R. Kachule, and J. Kotsopoulos. 2016. "The Quiet Rise of Medium-Scale Farms in Malawi." Land 5 (3): 19.

Arndt, C., K. Pauw, and J. Thurlow. 2016. "The Economy-wide Impacts and Risks of Malawi's Farm Input Subsidy Programme." American Journal of Agricultural Economics 98 (3): 962–980.

Babu, S., A. Comstock, B. Baulch, A. Gondwe, C. Kazembe, K. Kalagho, N.-L. Aberman, P. Fang, O. P. Mgemezulu, and T. Benson. 2018. "Assessment of the 2016/17 Food Insecurity Response Programme in Malawi." IFPRI Discussion Paper 01713, International Food Policy Research Institute, Washington, DC.

Barrett, C. B. 2008. "Smallholder Market Participation: Concepts and Evidence from Eastern and Southern Africa." Food Policy 33:299–317.

Barrett, C. B., L. Christiaensen, M. Sheahan, and A. Shimeles. 2017. "On the Structural Transformation of Rural Africa." Journal of African Economies 26 (Suppl. 1): i11–i35.

Baulch, B., and R. Botha. 2020. Can a Maize Price Band Work in Malawi? Malawi Strategy Support Program Policy Note 38. Lilongwe: International Food Policy Research Institute.

Beegle, K., A. Coudouel, and E. Monsalve. 2018. Realizing the Full Potential of Social Safety Nets in Africa. Africa Development Forum series. Washington, DC: World Bank.

Benfica, R., and J. Thurlow. 2017. Identifying Priority Value-Chains in Malawi. Rome and Washington, DC: International Fund for Agricultural Development and International Food Policy Research Institute.

Benin, S., J. Thurlow, X. Diao, C. McCool, and F. Simtowe. 2008. "Agricultural Growth and Investment Options for Poverty Reduction in Malawi." IFPRI Discussion Paper 00794, International Food Policy Research Institute, Washington, DC.

Benin, S., and A. Nin-Pratt. 2016. "Intertemporal Trends in Agricultural Productivity." In Agricultural Productivity in Africa: Trends, Patterns, and Determinants, edited by S. Benin, 25–104. Washington, DC: International Food Policy Research Institute.

Benson, T. 1999. Validating and Strengthening the Area-Specific Fertilizer Recommendations for Hybrid Maize Grown by Malawian Smallholders. Research report on nationwide 1997/98 maize fertilizer recommendations demonstration by Action Group I, Maize Productivity Task Force. Lilongwe: Ministry of Agriculture and Irrigation.

———. 2015. "Association between Irrigated Farming and Improved Nutrition in Farm Households in Malawi." Agrekon 54 (3): 62–86.

Benson, T., and F. Hartley. 2020. "Patterns of Change in Malawi's Economy under Sector-Focused Investment Strategies: Results of Scenarios Run through 2030 Using an Economy-wide Model for Malawi." Malawi Strategy Support Program Working Paper 32, International Food Policy Research Institute, Lilongwe.

Benson, T., A. Mabiso, and F. Nankhuni. 2016. Detailed Crop Suitability Maps and an Agricultural Zonation Scheme for Malawi: Spatial Information for Agricultural Planning Purposes. Feed the Future Innovation Lab for Food Security Policy Research Paper 17. East Lansing, US: Department of Agricultural, Food, and Resource Economics, Michigan State University.

Berge, E., D. Kambewa, A. Munthali, and H. Wiig. 2014. "Lineage and Land Reforms in Malawi: Do Matrilineal and Patrilineal Landholding Systems Represent a Problem for Land Reforms in Malawi?" Land Use Policy 41: 61–69.

Bezner-Kerr, R., S. Snapp, M. Chirwa, L. Shumba, and R. Msachi. 2007. "Participatory Research on Legume Diversification with Malawian Smallholder Farmers for Improved Human Nutrition and Soil Fertility." Experimental Agriculture 43 (4): 437–453.

Bhoojedhur, S., and T. Isbell. 2019. Almost Half of Malawians Consider Emigration: Most-Educated Are Most Likely to Look Overseas. Afrobarometer Dispatch no. 281. Accra: Afrobarometer.

Biscaye, P., C. L. Anderson, and T. Reynolds. 2017. Tracking Smallholder Farm Households. EPAR Technical Report 356. Seattle: Evans School Policy Analysis & Research Group, Daniel J. Evans School of Public Policy and Governance, University of Washington.

Blimpo, M. P., and M. Cosgrove-Davies. 2019. Electricity Access in Sub-Saharan Africa: Uptake, Reliability, and Complementary Factors for Economic Impact. Washington, DC: World Bank.

Bolton, A., and A. J. Bennett. 1975. Maize Breeding and Agronomy, Malawi 1971 to 1974. General Report, ODM Research Project R.2415. Lilongwe: Chitedze Agricultural Research Station.

Briggs, A. H., M. C. Weinstein, E. A. L. Fenwick, J. Karnon, M. J. Sculpher, and A. D. Paltiel. 2012. "Model Parameter Estimation and Uncertainty: A Report of the ISPOR-SMDM Modeling Good Research Practices Task Force-6." Value in Health 15 (6): 835–842.

Brown, P. 1966. "Maize Growing in Nyasaland. II. Fertilizer." Experimental Agriculture 2:49–60.

Burke, W. J., T. S. Jayne, and N. J. Sitko. 2020. "Do Medium-Scale Farms Improve Market Access Conditions for Zambian Smallholders?" Journal of Agricultural Economics 71:517–533.

Burke, W. J., and R. J. Myers. 2014. "Spatial Equilibrium and Price Transmission between Southern African Maize Markets Connected by Informal Trade." Food Policy 49:59–70.

Cardell, L., and H. Michelson. 2020. "Sell Low, Buy High?"—A New Explanation for a Persistent Puzzle. Department of Agricultural and Consumer Economics Working Paper. Champaign/Urbana, US: University of Illinois.

Carr, S. 2014. The Challenge of Africa's Nitrogen Drought: Some Indicators from the Malawian Experience. Malawi Strategy Support Program Policy Note 19. Lilongwe: International Food Policy Research Institute.

Chagunda, G., T. Gondwe, L. Banda, P. Mayuni, J. Mtimuni, T. Chimbaza, and A. Nkwanda. 2010. Smallholder Dairy Production in Malawi: Current Status and Future Solutions. Scoping paper for Optimising Smallholder Dairying project. Edinburgh: Scottish Agricultural College.

Chapoto, A., and T. S. Jayne. 2009. The Impacts of Trade Barriers and Market Interventions on Maize Price Predictability: Evidence from Eastern and Southern Africa. MSU International Development Working Paper 102. East Lansing, US: Michigan State University.

Chikowo, R., V. Chimonyo, C. Gwenambira, and S. Snapp. 2020. "Ecosystem Services in Doubled-Up Legume Systems." In The Role of Ecosystem Services in Sustainable Food Systems, edited by L. Rusinamhodzi, 171–180. London: Academic Press.

Chilowa, W. 1998. "The Impact of Agricultural Liberalisation on Food Security in Malawi." Food Policy 23 (6): 553–569.

Chimonyo, V. G. P., S. S. Snapp, and R. Chikowo. 2019. "Grain Legumes Increase Yield Stability in Maize Based Cropping Systems." Crop Science 59:1222–1235.

Chinseu, E. L., L. C. Stringer, and A. J. Dougill. 2019. "An Empirically Derived Conceptual Framework to Assess Disadoption of Conservation Agriculture: Multiple Drivers and Institutional Deficiencies." Journal of Sustainable Development 12 (5).

Chinsinga, B. 2015. "ADMARC in Contemporary Malawi: A Political Economy Analysis." Paper presented at OXFAM-Malawi symposium "Access to Agriculture Markets for Smallholder Farmers," Lilongwe, March 6.

———. 2016. "The Green Belt Initiative, Politics, and Sugar Production in Malawi." Journal of Southern African Studies 43 (3): 501–515.

Chinsinga, B., and C. Poulton. 2014. "Beyond Technocratic Debates: The Significance and Transience of Political Incentives in the Malawi Farm Input Subsidy Program." Development Policy Review 32 (S2): S123–S150.

Chirambo, M. 2016. "2015: When Malawi Tobacco Went Up in Smoke." The Nation (Malawi), January 7.

Chirwa, E. W. 2000. Food Marketing Reforms and Integration of Maize and Rice Markets in Malawi. Working Paper WC/05/99. Norwich, UK: School of Economics, University of East Anglia.

———. 2008. Land Tenure, Farm Investments and Food Production in Malawi. Research Programme Consortium on Improving Institutions for Pro-poor Growth Working Paper no. 18. Manchester: School of Environment and Development, University of Manchester.

Chirwa, E. W., and B. Chinsinga. 2013. Dealing with the 2007/08 Global Food Price Crisis: The Political Economy of Food Price Policy in Malawi. WIDER Working Paper no. 2013/30. Helsinki: UNU-WIDER.

Chirwa, E. W., and A. Dorward. 2013. Agricultural Input Subsidies: The Recent Malawi Experience. Oxford, UK: Oxford University Press.

Clapp, J. 2017. "Food Self-Sufficiency: Making Sense of It, and When It Makes Sense." Food Policy 66:88–96.

Corbeels, M., J. de Graaff, T. H. Ndah et al. 2014. "Understanding the Impact and Adoption of Conservation Agriculture in Africa: A Multi-scale Analysis." Agriculture, Ecosystems and Environment 187:155–170.

Coulibaly, J. Y., B. Chiputwa, T. Nakelse, and G. Kundhlande. 2017. "Adoption of Agroforestry and the Impact on Household Food Security among Farmers in Malawi." Agricultural Systems 155:52–69.

Darko, F. A., J. Ricker-Gilbert, T. Kilic, G. Shively, and R. Florax. 2016. Profitability of Fertilizer Use in SSA: Evidence from Malawi. Guiding Investment in Sustainable Agricultural Intensification in Africa Working Paper. West Lafayette, US: Purdue University.

Debalen, A., A. de la Fuente, A. Goyal, W. Karamba, N. T. V. Nguyen, and T. Tanaka. 2017. Pathways to Prosperity in Rural Malawi. Directions in Development series. Washington, DC: World Bank.

Deininger, K., and F. Xia. 2016. "Quantifying Spillover Effects from Large Land-Based Investment: The Case of Mozambique." World Development 87:227–241.

———. 2017. Assessing Effects of Large-Scale Land Transfers: Opportunities in Malawi's Estate Sector. Policy Research Working Paper 8200. Washington, DC: World Bank.

Deininger, K., F. Xia, and S. Holden. 2019. "Gendered Incidence and Impacts of Tenure Insecurity on Agricultural Performance in Malawi's Customary Tenure System." Journal of Development Studies 55 (4): 597–619.

Díaz-Bonilla, E. 2015. Macroeconomics, Agriculture, and Food Security: A Guide to Policy Analysis in Developing Countries. Washington, DC: International Food Policy Research Institute.

Dionne, K. Y., and J. Horowitz. 2016. "The Political Effects of Agricultural Subsidies in Africa: Evidence from Malawi." World Development 87:215–226.

Dorward, A., and J. Kydd. 2004. "The Malawi 2002 Food Crisis: The Rural Development Challenge." Journal of Modern African Studies 42 (3): 343–361.

Dorward, A., J. Kydd, C. Poulton, and D. Bezemer. 2009. "Coordination Risk and Cost Impacts on Economic Development in Poor Rural Areas." Journal of Development Studies 45 (7): 1093–1112.

Dorward, A. R. 2002. A Typology of Malawian Rural Households. Institutions and Economic Policies for Pro-poor Agricultural Growth working paper. London: Imperial College, Wye Campus.

———. 2006. "Markets and Pro-poor Agricultural Growth: Insights from Livelihood and Informal Rural Economy Models in Malawi." Agricultural Economics 35 (2): 157–169.

Douillet, M., and A. Toulon. 2014. "Developing a Typology of Agricultural Holdings for Improved Policy Design: A Preliminary Case Study of Malawi." FARM Working Paper 06. Fondation pour l'agriculture et la ruralité dans le monde, Montrouge, France.

Edelman, B., and N.-L. Aberman. 2015. Promoting Exports of Low-Aflatoxin Groundnut from Malawi. Malawi Strategy Support Program Policy Note 21. Lilongwe: International Food Policy Research Institute.

Edelman, B., A. Mabiso, Z. Nyirenda, and C. Kazembe. 2016. Have Market Policies Turned Malawi's Large-Scale Farmers into Subsistence Maize Producers? Malawi Strategy Support Program Policy Note 24. Lilongwe: International Food Policy Research Institute.

Ellis, F. 2005. "Small Farms, Livelihood Diversification, and Rural-Urban Transitions: Strategic Issues for Sub-Saharan Africa." Paper presented at research workshop, "The Future of Small Farms," organized by International Food Policy Research Institute, Overseas Development Institute, and Imperial College, London, Wye, UK, June 26–29.

Ellis, F., and E. Manda. 2012. "Seasonal Food Crises and Policy Responses: A Narrative Account of Three Food Security Crises in Malawi." World Development 40 (7): 1407–1417.

ELUS (Estate Lands Utilisation Study). 1997. Land, People, and Production on the Estates of Malawi—A Guide to the Surveys of the Estate Land Utilisation Study 1995–7. Lilongwe: Ministry of Lands and Valuation.

Elwell, H., and J. Rook. 1996. Soil and Water Conservation Technologies in Four Selected Countries in Southern Africa (Malawi, Mozambique, Zambia and Zimbabwe). Washington, DC: World Bank.

Eschweiler, J. A., S. Paris, J. H. Venema, A. J. M. Lorkeers, and R. I. Green. 1991. Methodology for Land Resources Survey and Land Suitability Appraisal. Land Resources Evaluation Project—Malawi Field Document No. 30. Lilongwe, New York, and Rome: Land Husbandry Branch, Malawi Ministry of Agriculture; United Nations Development Programme; and Food and Agriculture Organization of the United Nations.

Fafchamps, M. 1992. "Cash Crop Production, Food Price Volatility, and Rural Market Integration in the Third World." American Journal of Agricultural Economics 74 (1): 90–99.

FAO (Food and Agriculture Organization of the United Nations). 1996. Rome Declaration on World Food Security and World Food Summit Plan of Action. Rome.

———. 2015a. Analysis of Public Expenditure in Support of Food and Agriculture in Malawi. Monitoring and Analysing Food and Agricultural Policies (MAFAP) Technical Notes Series. Rome.

———. 2015b. Review of Food and Agricultural Policies in Malawi—Country Report 2014. Monitoring and Analysing Food and Agricultural Policies (MAFAP) Country Report Series. Rome.

———. 2020. FAOSTAT database. Accessed October. http://faostat3.fao.org/.

FEWS NET (Famine Early Warning Systems Network). 2012. Informal Cross Border Food Trade in Southern Africa. Issue 78. Pretoria: FEWS NET Regional Office.

———. 2016. Malawi Livelihood Baseline Profiles. Washington, DC.

———. 2017. Tanzania Remote Monitoring Update April 2017. Washington, DC.

———. 2018. Malawi—Enhanced Market Analysis. September. Washington, DC.

Giertz, Å., J. Caballero, M. Dileva, D. Galperin, and T. Johnson. 2015. Managing Agricultural Risk for Growth and Food Security in Malawi. Agricultural Global Practice Note 15. Washington, DC: World Bank.

Gilbert, R., T. Benson, and O. Ecker. 2019. Shifting Plates? Assessing Changes in Nutrient Availability and Dietary Patterns in Malawi: Household-Level Evidence from 2010/11 and 2016/17. Malawi Strategy Support Program Working Paper 30. International Food Policy Research Institute, Lilongwe.

Gilbert, R. A. 2004. "Best-Bet Legumes for Smallholder Maize-Based Cropping Systems of Malawi." In Green Manure/Cover Crop Systems of Smallholder Farmers, edited by M. Eilittä, J. Mureithi, and R. Derpsch, 153–174. Dordrecht, Netherlands: Springer.

Giller, K. E., E. Witter, M. Corbeels, and P. Tittonell. 2009. "Conservation Agriculture and Smallholder Farming in Africa: The Heretics' View." Field Crops Research 114 (1): 23–34.

Gollin, D., S. Parente, and R. Rogerson. 2002. "The Role of Agriculture in Development." American Economic Review 92 (2): 160–164.

Gowshall, M., and S. D. Taylor-Robinson. 2018. "The Increasing Prevalence of Non-communicable Diseases in Low-Middle Income Countries: The View from Malawi." International Journal of General Medicine 11:255–264.

Green, E. 2012a. "Labor Costs and the Failed Support of Progressive Farmers in Colonial Malawi." In Landscape, Environment, and Technology in Colonial and Postcolonial Africa, edited by T. Falola and E. Brownwell, 173–201. New York and London: Routledge.

———. 2012b. Land Concentration, Institutional Control and African Agency: Growth and Stagnation of European Tobacco Farming in Shire Highlands, c 1900–1940. Lund Papers in Economic History No. 125. Lund, Sweden: Department of Economic History, Lund University.

Grist, N. 2015. Case Study: Malawi's Agriculture, Climate Change and Food Security—Country Analysis and Programming Recommendations. Document for Department for International Development (DfID) Adaptation Knowledge and Tools program. London: Overseas Development Institute.

Gross, R., H. Schoeneberger, H. Pfeifer, and H.-J. Preuss. 2000. "The Four Dimensions of Food and Nutrition Security: Definitions and Concepts." SCN News 20:20–25.

Haggblade, S., P. B. R. Hazell, and P. A. Dorosh. 2007. "Sectoral Growth Linkages between Agriculture and the Rural Nonfarm Economy." In Transforming the Rural Nonfarm Economy: Opportunities and Threats in the Developing World, edited by S. Haggblade, P. B. R. Hazell, and T. A. Reardon, 141–182. Baltimore: Johns Hopkins University Press.

Haggblade, S., P. B. R. Hazell, and T. A. Reardon. 2007. Transforming the Rural Nonfarm Economy: Opportunities and Threats in the Developing World. Baltimore: Johns Hopkins University Press.

Haggblade, S., and B. Zulu. 2003. "The Recent Cassava Surge in Zambia and Malawi." Paper presented at the International Conference on Successes in African Agriculture, Pretoria, December 1–3.

Harrigan, J. 2001. From Dictatorship to Democracy: Economic Policy in Malawi 1964–2000. Aldershot, UK: Ashgate.

———. 2003. "U-Turns and Full Circles: Two Decades of Agricultural Reform in Malawi 1981–2000." World Development 31 (5): 847–863.

Hazell, P. B. R. 2017. "Why an Inclusive Agricultural Transformation Is Africa's Way Forward." In Africa Agriculture Status Report: The Business of Smallholder Agriculture in Sub-Saharan Africa, 1–19. Nairobi: Alliance for a Green Revolution in Africa.

Hazell, P. B. R., S. Wood, M. Bacou, and B. Dey. 2017. "Operationalizing the Typology of Small Farm Households." In Africa Agriculture Status Report: The Business of Smallholder Agriculture in Sub-Saharan Africa, 20–24. Nairobi: Alliance for a Green Revolution in Africa.

Henao, J., and C. Baanante. 1999. Estimating Rates of Nutrient Depletion in Soils of Agricultural Lands of Africa. Technical Bulletin IFDC-T48. Muscle Shoals, US: International Fertilizer Development Center.

Ikerra, S. T., J. A. Maghembe, P. C. Smithson, and R. J. Buresh. 2001. "Dry-Season Sesbania Fallows and Their Influence on Nitrogen Availability and Maize Yields in Malawi." Agroforestry Systems 52:13–21.

Inocencio, A., M. Kikuchi, M. Tonosaki, A. Maruyama, D. Merrey, H. Sally, and I. de Jong. 2007. Costs and Performance of Irrigation Projects: A Comparison of Sub-Saharan Africa and Other Developing Regions. IWMI Research Report 109. Colombo, Sri Lanka: International Water Management Institute.

IOM (International Organization for Migration). 2015. Migration in Malawi: A Country Profile 2014. Geneva.

Jayne, T. S., N. M. Mason, W. J. Burke, and J. Ariga. 2018. "Taking Stock of Africa's Second-Generation Agricultural Input Subsidy Programs." Food Policy 75:1–14.

Jayne, T. S., S. Snapp, F. Place, and N. Sitko. 2019. "Sustainable Agricultural Intensification in an Era of Rural Transformation in Africa." Global Food Security 20:105–113.

Jew, E. K. K., S. Whitfield, A. J. Dougill, D. D. Mkwambisi, and P. Steward. 2020. "Farming Systems and Conservation Agriculture: Technology, Structures and Agency in Malawi." Land Use Policy 95:104612.

Johnston, B., and J. Mellor. 1961. "The Role of Agriculture in Economic Development." American Economic Review 51 (4): 566–593.

Jones, R. B., J. W. Wendt, W. T. Bunderson, and O. A. Itimu. 1996. "Leucaena + Maize Alley Cropping in Malawi. Part 1: Effects of N, P, and Leaf Application on Maize Yields and Soil Properties." Agroforestry Systems 33:281–294.

Kalinga, O. J. M. 1993. "The Master Farmers' Scheme in Nyasaland, 1950–1962: A Study of a Failed Attempt to Create a 'Yeoman' Class." African Affairs 92:367–387.

Kalipeni, E. 1997. "Contained Urban Growth in Post-independence Malawi." East African Geographical Review, 19 (2): 49–66.

Keita, N. 2003. "Root and Tuber Crops: Concepts and Methods Recommended by FAO and Operational Issues." In Proceedings of the Expert Consultation on Root Crop Statistics. Volume II: Invited Papers. Paper 13. Rome: Food and Agriculture Organization of the United Nations.

Kettlewell, R. W. 1955. An Outline of Agrarian Problems and Policy in Nyasaland. Prepared for Legislative Council. Zomba: Government Printer.

———. 1965. "Agricultural Change in Nyasaland, 1945–1960." Food Research Institute Studies 5 (3): 229–285.

Keyser, J., and H. Tchale. 2010. Quantitative Value Chain Analysis. Policy Research Working Paper 5242. Washington, DC: World Bank.

Kim, E. K., and N. V. Loayza. 2019. Productivity Growth: Patterns and Determinants across the World. Policy Research Working Paper 8852. Washington, DC: World Bank.

Kopper, S. A., T. S. Jayne, and S. S. Snapp. 2020. "Sifting through the Weeds: Understanding Heterogeneity in Fertilizer and Labor Response in Central Malawi." Ecological Economics 169:106561.

Kydd, J. 1985. "Malawi in the 1970s: Development Policies and Economic Change." In Malawi: An Alternative Pattern of Development: Proceedings of a Seminar Held in the Centre of African Studies, University of Edinburgh, 24 and 25 May 1984, 295–380. Edinburgh: University of Edinburgh.

Kydd, J., and R. Christiansen. 1982. "Structural Change in Malawi since Independence: Consequences of a Development Strategy Based on Large-Scale Agriculture." World Development 10 (5): 355–375.

Lunduka, R., J. Ricker-Gilbert, and M. Fisher. 2013. "What Are the Farm-Level Impacts of Malawi's Farm Input Subsidy Program? A Critical Review." Agricultural Economics 44:563–579.

Malawi, MoAFS (Ministry of Agriculture and Food Security). 2012. Guide to Agricultural Production and Natural Resources Management in Malawi , rev. ed. Lilongwe: Agricultural Communications Branch, Department of Agricultural Extension Services, MoAFS.

Malawi, MoAIWD (Ministry of Agriculture, Irrigation and Water Development). 2016a. Contract Farming Strategy. Lilongwe.

———. 2016b. National Agriculture Policy. Lilongwe.

———. 2018. National Agricultural Investment Plan (NAIP): Prioritised and Coordinated Agricultural Transformation Plan for Malawi: FY 2017/18–2022/23. Lilongwe.

Malawi, MoIT (Ministry of Industry and Trade). 2012. Malawi National Export Strategy, 2013–2018. Lilongwe.

———. 2020. Malawi National Export Strategy II, 2020–2025. Lilongwe.

Malawi, NPC (National Planning Commission). 2020. Malawi's Vision: An Inclusively Wealthy and Self-Reliant Nation, Malawi 2063. Lilongwe.

Malawi, NSO (National Statistical Office). 2010. The 2006/07 Malawi National Census of Agriculture and Livestock. Zomba.

———. 2012. Integrated Household Survey, 2010–2011: Household Socio-economic Characteristics Report. Zomba.

———. 2014a. Integrated Household Panel Survey 2010–2013: Household Socio-economic Characteristics Report. Zomba.

———. 2014b. Malawi Labour Force Survey 2013. Zomba.

———. 2015. Welfare Monitoring Survey 2014. Report. Zomba.

———. 2017. Integrated Household Survey, 2016–2017: Household Socio-economic Characteristics Report. Zomba.

———. 2018. 2018 Population and Housing Census: Preliminary Report. Zomba.

———. 2020. "Trade Map—Malawi." Accessed October 2020. http://www.nsomalawi.mw.

Malawi, NSO (National Statistical Office) and ICF Macro. 2011. Malawi Demographic and Health Survey 2010. Zomba, Malawi, and Calverton, MD, US.

——— 2017. Malawi Demographic and Health Survey 2015–16. Zomba, Malawi, and Rockville, MD, US.

Malawi, NSO and World Bank. 2007a. “Malawi Poverty and Vulnerability Assessment: Investing in Our Future—Full Report.” Zomba and Washington, DC: NSO and World Bank.

———. 2007b. “Malawi Poverty and Vulnerability Assessment: Investing in Our Future—Appendices.” Zomba and Washington, DC: NSO and World Bank.

———. 2018. Methodology for Poverty Measurement in Malawi (2016/17). Zomba: NSO and World Bank Poverty and Equity Global Practice.

Mapemba, L. D. 1997. The Abandonment of Estates: Report of the Exploratory Study of Abandoned Estates 1997. Estate Land Utilisation Study. Lilongwe: Ministry of Lands and Valuation.

Masanjala, W. H. 2018. Brain Drain in Africa: The Case of Tackling Capacity Issues in Malawi’s Medical Migration. ACBF Occasional Paper no. 31. Harare: African Capacity Building Foundation.

Mayén, A-L., P. Marques-Vidal, F. Paccaud, P. Bovet, and S. Stringhini. 2014. “Socioeconomic Determinants of Dietary Patterns in Low- and Middle-Income Countries: A Systematic Review.” The American Journal of Clinical Nutrition 100 (6): 1520–1531.

McMillan, M. S., D. Rodrik, and C. Sepúlveda. 2017. “Overview: Structural Change, Fundamentals, and Growth.” In Structural Change, Fundamentals, and Growth: A Framework and Case Studies, edited by M. S. McMillan, D. Rodrik, and C. Sepúlveda, 1–38. Washington, DC: International Food Policy Research Institute.

Mellor, J. W. 2014. “High Rural Population Density in Africa: What Are the Growth Requirements and Who Participates?” Food Policy 48:66–75.

———. 2017. Agricultural Development and Economic Transformation: Promoting Growth with Poverty Reduction. Cham, Switzerland: Palgrave Macmillan.

Mellor, J. W., and S. Malik. 2017. “The Impact of Growth in Small Commercial Farm Productivity on Rural Poverty Reduction.” World Development 91:1–10.

MERA (Malawi Energy Regulatory Authority). 2017. Malawi Renewable Energy Strategy. Lilongwe.

———. 2019. Regulatory Framework for Mini-grids. Lilongwe.

Minot, N. 2014. “Food Price Volatility in Sub-Saharan Africa: Has It Really Increased?” Food Policy 45:45–56.

Morris, M. L., J. Risopoulos, and D. Beck. 1999. Genetic Change in Farmer-Recycled Maize Seed: A Review of the Evidence. CIMMYT Economics Working Paper No. 99-07. El Batán, Texcoco, Mexico: CIMMYT.

Moyer-Lee, J., and M. Prowse. 2015. "How Traceability Is Restructuring Malawi's Tobacco Industry." Development Policy Review 33:159–174.

Mueller, V., A. Quisumbing, H. L. Lee, and K. Droppelmann. 2014. "Resettlement for Food Security's Sake: Insights from a Malawi Land Reform Project." Land Economics 90 (2): 222–236.

Murray, S. 2017. New Technologies Create Opportunities. WIDER Working Paper 2017/156. Helsinki: UNU-WIDER.

Mussa, R. 2015. "Do the Poor Pay More for Maize in Malawi?" Journal of International Development 27:546-563.

Myers, R. J. 2013. "Evaluating the Effectiveness of Inter-regional Trade and Storage in Malawi's Private Sector Maize Markets." Food Policy 41:75–84.

NAFSN (New Alliance for Food Security and Nutrition). 2013. Country Cooperation Framework to Support the New Alliance for Food Security and Nutrition in Malawi. G8 New Alliance for Food Security and Nutrition.

Ng'ong'ola, C. 1986. "Malawi's Agricultural Economy and the Evolution of Legislation on the Production and Marketing of Peasant Economic Crops." Journal of Southern African Studies 12 (2): 240–262.

Nkhoma, P. R. 2018. "The Evolution of Agricultural Input Subsidy Programs: Contextualizing Policy Debates in Malawi's FISP." World Development Perspectives 9:12–17.

Nkhoma, P. R., M. M. Bosman, and M. Eduful. 2019. "Constituting Agricultural and Food Security Policy in Malawi: Exploring the Factors That Have Driven Policy Processes in the Farm Inputs Subsidy Programme." Journal of Asian and African Studies 54 (3): 360–375.

Ochieng, D. O., R. Botha, and B. Baulch. 2019. "Structure, Conduct, and Performance of Maize Markets in Malawi." Malawi Strategy Support Program Working Paper 29, International Food Policy Research Institute, Lilongwe.

Otañez, M. G., H. Mamudu, and S. A. Glantz. 2007. "Global Leaf Companies Control the Tobacco Market in Malawi." Tobacco Control 16:261–269.

Pauw, K., P. Dorosh, and J. Mazunda. 2013. "Exchange Rate Policy and Devaluation in Malawi." IFPRI Discussion Paper 01253. International Food Policy Research Institute, Washington, DC.

Pauw, K., and B. Edelman. 2015. "Is Malawi's Mix of Maize Market Policies Ultimately Harming Food Security?" Malawi Strategy Support Program Policy Note 22. Lilongwe: International Food Policy Research Institute.

Peters, P. E. 2004. "Inequality and Social Conflict over Land in Africa." Journal of Agrarian Change 4 (3): 269–314.

Petersen, B., and S. Snapp. 2015. "What Is Sustainable Intensification? Views from Experts." Land Use Policy 46:1–10.

Phiri, R. H., S. Snapp, and G. Y. Kanyama-Phiri. 1999. "Soil Nitrate Dynamics in Relation to Nitrogen Source and Landscape Position in Malawi." Agroforestry Systems 47:253–262.

Popkin, B. M., L. Adair, and S. W. Ng. 2012. "Global Nutrition Transition and the Pandemic of Obesity in Developing Countries." Nutrition Reviews 70 (1): 3–21.

Potts, D. 2006. "Rural Mobility as a Response to Land Shortages: The Case of Malawi." Population, Space and Place 12:291–311.

Pretty, J., and Z. P. Bharucha. 2014. "Sustainable Intensification in Agricultural Systems." Annals of Botany 114:1571–1596.

Price, A. J., A. C. Crampin, A. Amberbir et al. 2018. "Prevalence of Obesity, Hypertension, and Diabetes, and Cascade of Care in Sub-Saharan Africa: A Cross-Sectional, Population-Based Study in Rural and Urban Malawi." Lancet Diabetes & Endocrinology 6 (3): 208–222.

Prowse, M. 2013. "A History of Tobacco Production and Marketing in Malawi, 1890–2010." Journal of Eastern African Studies 7 (4): 691–712.

Prowse, M., and P. Grassin. 2020. Tobacco, Transformation and Development Dilemmas from Central Africa. Cham, Switzerland: Palgrave Macmillan.

Prowse, M., and J. Moyer-Lee. 2014. "A Comparative Value Chain Analysis of Smallholder Burley Tobacco Production in Malawi—2003/4 and 2009/10." Journal of Agrarian Change 14 (3): 323–346.

Pryor, F. L., and C. Chipeta. 1990. "Economic Development through Estate Agriculture: The Case of Malawi." Canadian Journal of African Studies 24 (1): 50–74.

Ricker-Gilbert, J., C. B. Jumbe, J. Kanyamuka, J. Chamberlin, R. Lunduka, and S. Kaiyatsa. 2019. "How Do Informal Farmland Rental Markets Affect Smallholders' Well-Being? Evidence from a Matched Tenant-Landlord Survey in Malawi." Agricultural Economics 50 (5): 595–613.

Robertson, M. J., W. Sakala, T. Benson, and Z. Shamudzarira. 2005. "Simulating Response of Maize to Previous Velvet Bean (Mucuna pruriens) Crop and Nitrogen Fertiliser in Malawi." Field Crops Research 91 (1): 91–105.

Royal Society. 2009. Reaping the Benefits: Science and the Sustainable Intensification of Global Agriculture. London.

Sahn, D., J. Arulpragasam, and L. Merid. 1990. Policy Reform and Poverty in Malawi: A Survey of a Decade of Experience. Cornell Food and Nutrition Policy Program Monograph 7. Ithaca, US: Cornell University, Division of Nutritional Sciences.

Saka, A. R., W. T. Bunderson, O. A. Itimu, H. K. S. Phombeya, and Y. Mbekeyani. 1994. "The Effects of Acacia albida on Soils and Maize Grain Yields under Smallholder Farm Conditions in Malawi." Forest Ecology and Management 64 (2–3): 217–230.

Saka, J. D. K., P. Sibale, T. S. Thomas, S. Hachigonta, and L. M. Sibanda. 2013. "Malawi." In Southern African Agriculture and Climate Change: A Comprehensive Analysis, edited by S. Hachigonta, G. C. Nelson, T. S. Thomas, and L. M. Sibanda, 111–146. Washington, DC: International Food Policy Research Institute.

Sen, A. 1981. Poverty and Famines: An Essay on Entitlement and Deprivation. London: Clarendon.

Shekar, M., and B. Popkin, eds. 2020. Obesity: Health and Economic Consequences of an Impending Global Challenge. Washington, DC: World Bank.

SMEC (Snowy Mountains Engineering Corporation International Pty Ltd). 2015a. National Irrigation Master Plan and Investment Framework—Appendix 10: Financial and Economic Analysis. Lilongwe: Ministry of Agriculture, Irrigation, and Water Development, Department of Irrigation.

———. 2015b. National Irrigation Master Plan and Investment Framework—Main Report. Lilongwe: Ministry of Agriculture, Irrigation, and Water Development, Department of Irrigation.

Smith, A., S. Snapp, J. Dimes, C. Gwenambira, and R. Chikowo. 2016. "Doubled-Up Legume Rotations Improve Soil Fertility and Maintain Productivity under Variable Conditions in Maize-Based Cropping Systems in Malawi." Agricultural Systems 145:139–149.

Snapp, S., T. S. Jayne, W. Mhango, T. Benson, and J. Ricker-Gilbert. 2014. "Maize Yield Response to Nitrogen in Malawi's Smallholder Production Systems." Malawi Strategy Support Program Working Paper 06. International Food Policy Research Institute, Lilongwe.

Steele, R. J. G. 1997. Listing Survey 1995: Final Report. Lilongwe: Malawi Estate Land Utilisation Study, Ministry of Lands and Valuation.

Swindale, A., and P. Bilinsky. 2006. Household Dietary Diversity Score (HDDS) for Measurement of Household Food Access: Indicator Guide (v.2). Washington, DC: FHI 360/FANTA.

Taulo, J. L., K. J. Gondwe, and A. B. Sebitosi. 2015. "Energy Supply in Malawi: Options and Issues." Journal of Energy in Southern Africa 26 (2): 19–32.

Thierfelder, C., P. Chivenge, W. Mupangwa, T. S. Rosenstock, C. Lamanna, and J. X. Eyre. 2017. "How Climate-Smart Is Conservation Agriculture (CA)? Its Potential to Deliver on Adaptation, Mitigation and Productivity on Smallholder Farms in Southern Africa." Food Security 9:537–560.

Thurlow, J. 2017. 2014 Social Accounting Matrix for Malawi: A Nexus Project SAM. Washington, DC: International Food Policy Research Institute.

Timmer, P. 2015. Food Security and Scarcity: Why Ending Hunger Is So Hard. Philadelphia: University of Pennsylvania Press.

Tittonell, P. 2014. "Ecological Intensification of Agriculture—Sustainable by Nature." Current Opinion in Environmental Sustainability 8:53–61.

UNICEF. 1990. Strategy for Improved Nutrition of Children and Women in Developing Countries: A UNICEF Policy Review. New York.

USAID (United States Agency for International Development). 1992. Definition of Food Security. USAID Policy Determination 19. Washington, DC.

———. 2010. USAID Country Profile—Property Rights and Resource Governance: Malawi. Washington, DC: USAID.

Vail, L. 1983. "The State and the Creation of Colonial Malawi's Agricultural Economy." In Imperialism, Colonialism, and Hunger: East and Central Africa, edited by R. I. Rotberg, 39–87. Lexington, US: Lexington Books.

Vaughan, M. 1985. "The Politics of Food Supply: Colonial Malawi in the 1940s." In Malawi: An Alternative Pattern of Development: Proceedings of a Seminar Held in the Centre of African Studies, University of Edinburgh, 24 and 25 May 1984, 67–91. Edinburgh: University of Edinburgh.

———. 1987. The Story of an African Famine: Gender and Famine in Twentieth-Century Malawi. Cambridge: Cambridge University Press.

WFP (World Food Programme). 2014. "Protracted Relief and Recovery Operations—Malawi 200692." Project document for WFP executive board approval, second regular session, November 10–13.

Whiteside, M. 2003. Enhancing the Role of Informal Maize Imports in Malawi's Food Security. Consultancy report for the Department for International Development (DfID), United Kingdom. Lilongwe: DfID–Malawi.

Wineman, A., T. Njagi, C. L. Anderson, T. W. Reynolds, P. Wainaina, E. Njue, P. Biscaye, and M. W. Ayieko. 2019. A Case of Mistaken Identity? Measuring Rates of Improved Seed Adoption in Tanzania Using DNA Fingerprinting. EPAR Technical Report 363. Seattle: Evans School Policy Analysis and Research, Daniel J. Evans School of Public Policy and Governance, University of Washington.

World Bank. 2013. World Development Report 2014: Risk and Opportunity—Managing Risk for Development. Washington, DC.

———. 2018. Malawi Economic Monitor—Realizing Safety Nets' Potential. Macroeconomics and Fiscal Management Global Practice. Washington, DC.

———. 2019. Malawi Economic Monitor—Charting a New Course. Macroeconomics and Fiscal Management Global Practice. Washington, DC.

———. 2020. World Development Indicators dataset. Accessed October. https://data.worldbank.org/.

Author

Todd Benson is a senior research fellow in the Development Strategy and Governance Division of the International Food Policy Research Institute, Washington, DC.

Made in the USA
Las Vegas, NV
03 November 2026

57423043R00121